Ab initio Calculation Tutorial

Ryo Maezono

Ab initio Calculation Tutorial

For Materials Analysis, Informatics and Design

 Springer

Ryo Maezono 🅾
School of Information Science
Japan Advanced Institute of Science
and Technology
Nomi, Ishikawa, Japan

ISBN 978-981-99-0921-6   ISBN 978-981-99-0919-3   (eBook)
https://doi.org/10.1007/978-981-99-0919-3

This Springer imprint is published by the registered company Springer Nature Singapore Pte Ltd.
The registered company address is: 152 Beach Road, #21-01/04 Gateway East, Singapore 189721,
Singapore

Preface

Electronic structure calculations have become mature enough in method, and many packages are now available and well maintained for practitioners to use them conveniently. It has already left the hands of theoreticians, being widely spreading over practical domains including industries and experimental groups. Concepts such as "material genome" or "material informatics" as a fusion with machine learning and artificial intelligence to realize novel materials search comprehensively have also become a popular word and are rapidly being introduced in industry and experimental laboratories. Group leaders in such non-major fields are beginning to think that they want their young members and new students to acquire such computational framework, even though they themselves are not majors in simulations. Through many collaborations with experimental group that does not major in simulation, the author has gained experience of constructing effective teaching programs for technology transfer during the short stays of students sent from them staying in my group.

I have heard many greetings from the group leaders who sent students such as "I have tried to read through several textbooks on my own, but I gave up to found it difficult to understand 'the main line' of what is explained. That's why I decided to ask for your guidance". Putting all these comments together, I found that most of the courses designed for beginners seem to be structured based on the purpose of the application such as "calculate surface catalytic properties, optical absorption, thermal conductivity, ...", *etc.* The courses are taking a specific software package appropriate to each purpose, and then provide "a training session" on how to use each package. What they are whinging seems that "Although I understood each line of the explanation, I could not see the background concept, which gave me an unclear feeling".

Not only in textbooks, but also in tutorial workshops for beginners, the contents tend to be juxtaposition of methodologies such as density functional theory, molecular orbital method, molecular dynamics, GW, TDDFT.... This makes it difficult for beginners to understand the relationship and necessity of each method. To make matters worse, as the workshop year goes on, there is an overload of this and that, giving the impression that the content is too advanced and too much for beginners. Theoretical methodologies have evolved over the years, allowing for more realistic analysis of phenomena. Therefore, the instructors are likely to think, "It would be better to include these methods in the course to meet the demand of the participants", and so they pile up more and more advanced theoretical methodologies. Also, the participants may say, "I am working on this kind of phenomenon, and I have heard that 'the XX-method' can handle it, so I would like to participate in an opportunity to learn it". As a result, such an event is successful just in terms of the number of participants, but there is a limit to how much advanced methodology can be taught to beginners in a limited time, making participants "indigestion". If such a course were to be published in textbook, it would become a flat, enumerative procedure book, which would lose the pleasure of reading through it, and the reader's desire to learn would stall. If one decides to learn the background theory so that one can look over the whole picture rather than individual procedures, the beginners would turn to expert books on electronic structure calculations. However, these expert books are designed for theoretical researchers who are developing methodologies, and are not written from the perspective for beginners with a specific mission to become a practitioner as quickly as possible.

In fact, the author has own difficult experience regarding the above. After obtaining a Ph.D. majoring magnetic theory, I started my career in the field of electronic structure calculations as a postdoctoral researcher abroad. Mastering *ab initio* electronic structure calculations requires an amount of collective knowledge, spanning physics/chemistry, computer hardware, software and programming, Linux systems, and individual methodologies of electronic structure calculations, in addition to knowledge of many-body electronic theory. In such a "forest of knowledge", it is impossible for a beginner to identify what are the main paths and what are the side paths, and one easily gets lost the way not to understand the hierarchical construction of the knowledge. Practitioners new to the field do not have the luxury of taking the time to sort out the big picture, unlike graduate students who can spend 5 years slowly studying in a lab dedicated to simulation. It's like looking around a large museum, you need a guide who can walk through it with you in a few hours first, explaining where is the key spot and how each key relates each other. It is important to have the guide clearly indicate that this is a side road that can be visited later, and a few words such as "Well, you'll figure this out later", or "I didn't figure this out right away either", will reassure the beginner as they follow the itinerary.

This book is composed based on the contents of tutorial courses developed by the author's group for industrial engineers or researchers who are mainly engaged in experiments under the collaborations. Along the above-mentioned policy, the book is structured to avoid "the overload" to minimize the contents as least as possible. It is designed for readers to feel successful experience like "arriving at 'ridge road'" as quickly as possible rather than getting bored to follow long stories of background theories. For this purpose, it provides several practical operations appeared in earlier stage where the readers might feel to work on a black box. This is the opposite of the usual descriptive style, *i.e.*, a textbook structure that preaches from the basic theory and proceeds to practical training. In the first half of the book, the chapters focus solely on "what solutions the *ab initio* calculations provide", "what should be given as input, and what is obtained as output" from the practitioner's viewpoint. Detailed explanations of the background theory are given in the latter chapters in a way that makes the reader realize "the operations I experienced in the earlier chapters are actually correspond to this!". It is often the case for readers to lose motivation for learning if a textbook starts with individual details in order for many chapters without a bird's-eye view of the overall structure. This is just like the situation where pupils with not enough interests in history and culture take school trip to visit historical temples and shrines. Such a trip is much more effective to impress the contents for experienced adults (the author refers to as a "school trip for adults"). However, it is also true for students being difficult to understand what a teacher is talking about the overall summary first without knowing the individual topics at all. Therefore, even though there are redundancies in repeated appearances, the structure is such that the explanation of the basic theory reappears again, gradually deepening in depth several times, like "cutting a thick cardboard little by little with a cutter knife". For readers who have basic knowledge and do not require much hands-on practice, they can skip the contents up to Chap. 3.

One more feature different from preceding textbooks is that we emphasized the difference and interrelationship between "the kernel part" performing the *ab initio* calculation itself, and "the outer utilities" to perform the evaluation to get various properties realized by the outer loop to call the kernel. Down from the time when the preceding books introducing *ab initio* calculations were written, today there are conveniently maintained functions to evaluate several properties, those are surely achieved by "*ab initio*" packages but performed actually by "non-*ab initio*" calculations using model formulas. The author has come to realize that this is one of the factors that confuse beginners, through my experience to train industrial practitioners and experimental researchers. Such confusion can continue to cast a hazy shadow, which can have a significant impact on maintaining a motivation to learn. Therefore, we structured the content in such a way that the "hierarchical arrangement of the actors in the show" (the relation between "the kernel calculation" and "the outer evaluations") could be clearly understood at the beginning. We then clearly emphasized why we should focus on learning the kernel part only, so that readers can be convinced to work on. This is what the author was aware of as a distinctive role played by an introductory book published in this period on top of the preceding books.

Tutorial Environment

This book is designed as a self-study textbook to learn DFT through simulation practice. It is the philosophy of this book not to stall the learning motivation of beginners. Even if one provides a tutorial book based on the assumption that readers have already prepared an computational environment in a university laboratory, or that they have already learned Linux, it will be of no use for such a beginner who has only experience to use Windows PC to start one's self-study at home by oneself.

This book carefully explains the process of building a simulation environment for beginners to enjoy and learn how to operate a simulation using commands. It is not possible to take an unrealistic premise such as "First, prepare a Linux server" for such beginners who are unfamiliar with the term "Linux server". As such, we have designed and provided a course that explains how to build a simulation environment on a personal PC, such as a Win/Mac, with which beginners are familiar. As a new attempt adopted in this book, we have decided to provide a course using the "RaspberryPi", which can be considered an ultra-cheap Linux server (total cost is about 130 USD).

The "RaspberryPi" has been heard of in a variety of situations. This book may provide a good opportunity of it. There are probably many people who are interested in purchasing one because of its low price, but have been put off by the fact that they cannot find any use for it (as was the case with the author until writing this book). It would be impressive that a processor without a cooling fan is able to perform such simulations to solve differential equations. I hope this book will be a good first contact to "the world of computer architecture" to attract readers' interests of it.

Cost Estimation for Simulation Environment

In this book, a hands-on tutorial course is provided using "Raspberry Pi 4", for which the required expenses are estimated as follows (supposing a reader already possessing display/keyboard/mouse with Wi-Fi availability):

- Raspberry Pi 4 Model B (8gb) [~ 90 USD],
- SD card/32GB [~ 10 USD],
- Micro HDMI adopter [~ 10 USD],
- AC-adopter [~ 7 USD],
- SD card reader [~ 15 USD],

amounting to around 130 USD in total.

Asahidai, Nomi, Ishikawa, Japan   Ryo Maezono
January 2023

Acknowledgements

The content of this book is based on a Japanese book (ISBN-13: 978-4627170315) published by Morikita Publishing in 2020. The concept of this book was based on the tutorial training program funded by JST-Sakura Science plan, which accepted intern students from Asian countries, the tutorial training program ASESMA (African School on Electronic Structure Methods and Applications) funded by ICTP (International Centre for Theoretical Physics), the workshop at Department of Metallurgical and Materials Engineering of IITM (Indian Institute of Technology, Madras), at Department of Materials Science of Chulalongkorn University (Thailand), *etc.* In particular, I would like to express my respectful acknowledgement to Prof. Richard Martin for his guidance in these activities.

I would like to thank Prof. Richard Needs and Dr. Mike D. Towler, and other members of the TCM group in Cavendish Laboratory (University of Cambridge) during my staying period, who helped me land in this field as a beginner in *ab initio* electronic structure calculations.

Dr. Tom Ichibha and Dr. Genki I. Prayogo have significantly assisted to prepare the tutorial materials for this book. I am grateful to my longtime research partner Prof. Kenta Hongo, who was deeply involved in putting this concept together. I also thank Dr. Kousuke Nakano who is originally from the industrial domain and assisted the concept of tutorial course. I would also like to thank Ph.D. students in my group who helped to verify the contents via several workshops (Dr. Apichai Jomphoak, Dr. Ornin Srihakulung, Dr. Keishu Utimula, Dr. Adie Hanindriyo, Dr. Qin Ken, Dr. Abhishek Raghav, Mr. Peng Song, Mr. Kenji Oqmhula, Mr. Abdul Ghaffar, Mr. Rohit Dahule).

Contents

Acronyms

The abbreviations that appear frequently in this document are summarized as follows:

DFT   Density Functional Theory

This book often adopts the following description conventions:

- **Square brackets used to accommodate variable names** In such a description,

  ```
  % cp [file01] [file02]
  ```

  the square brackets [...] accommodate variable names. It would actually be

  ```
  % cp maezono hongo
  ```

  for instance, with "[file01] = maezono" and "[file01] = hongo". Do not enter as is written with the square brackets.

- **Notation for commands that are too long to fit on a single line** A long command

  ```
  [A too long statement impossible to be accommodated]
  ```

  is sometimes represented as

  ```
  [A too long statement impossible \ to be
  accommodated]
  ```

  in two lines using "\". The "\" should be interpreted as "delete me and make it into one line". The system, however, will recognize it correctly even if you type it in as is.

Part I

Tracing the Whole Picture First

In this part, we will first outline how *ab initio* calculations are used and how they have evolved toward more practical use recently (Chap. 1).

From the later part (after Chap. 4), we provide a detailed explanation of the background theory, but first we will trace a series of operations (Chap. 3) so that you can get a concrete image of "Oh, that's what those operations mean" in the later parts. Therefore, in this part, please proceed with the practical training, keeping in one's mind that the objective here is first understanding a series of work flow even without understanding detailed background.

Chapter 2 describes the technical matters required for preparing the tutorial environment used in Chap. 3. The appendix chapter, "A short course of Linux command operation (Chap. 9)", is derived from Chap. 2. The contents of the appendix is technical in nature, but it is important to learn it in order to avoid stalling your learning process. The benefits are not limited to this tutorial, but will also improve the efficiency of the reader's daily work.

In another Chap. 11 derived from Chap. 3, we have given another style of explanation on the band theory along the motivation to reply the question "why you want to know about band dispersion?"

Introduction

Abstract

Recently, we often hear the terms "materials informatics" used in a variety of contexts. It is talked about like a dream technology in which artificial intelligence finds new materials, but the kernel part of the technology is a rather boring black box that just calculates the energy value for a given arrangement of atomic positions. This is where the confusion arises for the beginners. Though they are motivated by hearing that it can simulate fascinating physical properties, what they find in the textbook is how to calculate energies and bands which never seem what they want to do (Fig. 1.1). The primary purpose of this chapter is to dispel the sense of bewilderment that reduces such motivation. We will first look at how the reader's "final goal" (metaphorically speaking, the ridge trail in mountain climbing) and the "initial things to learn" (the boring forest before getting to the ridge trail) are connected. The "boring things" are to learn the input specification to be given to the boring black box. We will give explanations that will allow the readers to tackle this course with ease/confidence from Chap. 2 onward. In addition, by explaining further goal to aim at after mastering this book, we describe why command line usage is necessary, narrowing down the technical policies employed in this book.

1.1 What Is ab initio Electronic Structure Calculation in Practice?

1.1.1 What Solution It Provides?

Let us call **nanomaterial development** meaning such developments of materials whose properties are tuned at the level of atomic structure, such as atomic arrangement in a material. For example, when trying to make a harder solid for polishing material, one can improve the hardness of the material by replacing some of the elements (e.g., carbon) that make up the conventional material with boron [1]. In the emergence of the Bronze Age culture, we found that a small amount of tin was added

© The Author(s), under exclusive license to Springer Nature Singapore Pte Ltd. 2023
R. Maezono, *Ab initio Calculation Tutorial*,
https://doi.org/10.1007/978-981-99-0919-3_1

1/ 'Output quantity of the simulation does NOT explain what we want to analyze directly...'

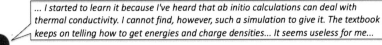

... I started to learn it because I've heard that ab initio calculations can deal with thermal conductivity. I cannot find, however, such a simulation to give it. The textbook keeps on telling how to get energies and charge densities... It seems useless for me...

→ Thermal conductivity is never the direct output of ab initio simulations. It is evaluated from the fitting coefficients over the dependence of the energies increased by given deformations (evaluation is achieved by the outer loop to perform single-point calculations repeatedly).

2/ 'Not a simulation as a whole description'

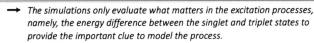

... I started to learn it because I've heard that ab initio calculations can handle the photo-excitation processes. Where is such a simulation to give the animations of the processes??

→ The simulations only evaluate what matters in the excitation processes, namely, the energy difference between the singlet and triplet states to provide the important clue to model the process.

Fig. 1.1 Typical cases where beginners get disappointed for ab initio calculations in practice. To avoid prematurely discouraging to learn, it is necessary to first comprehend what kind of solution it provides

to squishy, bendable copper, it became surprisingly hard and could be used to make swords. There has been such empirical knowledge that humans have developed as metallurgy since ancient times. Today, such knowledge is being understood further from the viewpoint of atomic level [2].

The same is true in the field of pharmaceuticals. There are pharmacological molecules that have the same elemental ratios, the same mass, melting point, and other properties, but physiologically, some of them have detrimental effects. We are now making progress in understanding these problems from the viewpoint of the steric structure of the molecule. Nowadays, research on manipulating the steric structure of such molecules at the atomic level and designing pharmacological agents has also been developed [3].

How to achieve low-cost and efficient chemical reactions for the industrial synthesis of useful materials has long been a very important research topic. The use of certain additives that do not themselves undergo chemical changes can magically increase the efficiency of reactions or promote reactions that would otherwise not proceed (catalytic effect). When the author was in high school, I heard about this and was impressed by the very magic of it, but this is now understood from atomic-level viewpoint that the molecular structure is modulated into a form that is easier to react when the molecules are brought closer to the catalyst [2,4]. Once this is understood, such research approach is possible to develop inexpensive catalysts so that the less expensive substance will deform the molecular structure of the reaction target in a more desirable way, from atomic-level viewpoint.

As you can see, the underlying common mechanism is simple although the subjects and fields of study are wide ranging, from inorganic to organic, metallurgy/chemistry/semiconductors/pharmacy/agricultural science. Most of the properties that are useful to mankind are almost determined by circumstances at the electronic level.[1] The situation at the electron level (**electronic structure**) is determined by given geometrical locations of atomic elements that make up the matter (**geometry**). Thus, we arrive at the idea of aggressively manipulating the atomic geometry to improve its properties.

We mentioned that the electronic structure is determined by giving the geometry. Fortunately, this procedure has been well established in the early twentieth century as a procedure to solve a quantum mechanical equation called the Schrodinger equation [5]. The ab initio electronic structure calculations covered in this book refer to simulations to solve the equation under the given geometry.

1.1.2   What Is Calculated?

An ab initio electronic structure "calculation" referred in this book is, specifically, a calculation to solve the equation

$$\left[-\frac{1}{2} \sum_{j=1}^{N} \nabla_j^2 + V\left(\mathbf{r}_1, \ldots, \mathbf{r}_N\right) \right] \cdot \Psi\left(\mathbf{r}_1, \ldots, \mathbf{r}_N\right) = E \cdot \Psi\left(\mathbf{r}_1, \ldots, \mathbf{r}_N\right) \quad (1.1)$$

numerically [2,4,5]. This equation (many-body Schrodinger equation) is explained in detail later in Sect. 5.3.1. "Solving the equation" means the task to find a "value E" and a "function $\Psi(\mathbf{r}_1, \ldots, \mathbf{r}_N)$" which satisfies the equality in the equation. In any application packages of ab initio electronic structure calculation, the numerical program code to solve the equation is set as their most kernel part just like "**reactor core**".

Depending on whether the target system is a semiconductor or a pharmacological molecule, the setting of **potential term** $V(\mathbf{r}_1, \ldots, \mathbf{r}_N)$ changes. The mission of a user is to provide an input file that instructs the program on these settings. If we focus only on the kernel part, all we can get from it is "the value E" and "the function $\Psi(\mathbf{r}_1, \ldots, \mathbf{r}_N)$", from which we can obtain "the energy value of the system" and the charge density. In other words, only the energy and charge density can be obtained from the **calculation**.

Knowing this, beginners may be disappointed, saying that "I took the course to learn ab initio calculations because I was told I could calculate thermal conductivity..." or "I was told I could calculate spectra...", etc. However, it is too early to close this book because the above "disappointing" statement is describing only about the kernel calculation. As explained in Fig. 1.2, the "single-point calculations"

---

[1] A typical exception is the use of nuclear energy. This comes from the situation at the level of nuclear reactions, which is a different hierarchy than the electron level. It was not until the twentieth century that nuclear energy was actively extracted and utilized, but as for energy at the electron level, it can be traced back to the use of fire. Combustion is nothing more than the release of energy that binds electrons together.

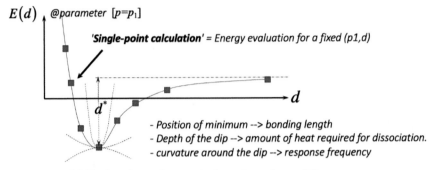

$E(d)$ @parameter $[p=p_1]$

'Single-point calculation' = Energy evaluation for a fixed (p1,d)

d^*

- Position of minimum --> bonding length
- Depth of the dip --> amount of heat required for dissociation.
- curvature around the dip --> response frequency

(a) '1st-layer' outer loop capturing a dependence of the energy.

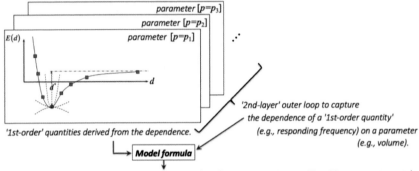

parameter $[p=p_3]$
parameter $[p=p_2]$
parameter $[p=p_1]$

$E(d)$

'1st-order' quantities derived from the dependence.

Model formula

'2nd-layer' outer loop to capture
the dependence of a '1st-order quantity'
(e.g., responding frequency) on a parameter
(e.g., volume).

'2nd-order' quantities (Conductivity, Expansion rate, Transition temperature, etc.)

(b) '2nd-layer' outer loop to evaluate '2nd-order' quantities.

Fig. 1.2 The relation between the single-point calculations (the kernel of ab initio package) and the outer loop to provide various physical properties. A single-point calculation just provides the energy value and charge density under a fixed condition. By calling this single-point calculation repeatedly with various conditions (e.g., lattice constants) under the loop structure of the programing, it is possible to evaluate the *dependence* of the energy value and charge density on the varying condition. From such dependence of energy value, several physical properties other than energy, such as response frequency, can be evaluated. Then the further outer loop calls the evaluation of the properties (e.g., phonon frequencies) repeatedly with varying the conditions (e.g., pressure, volume, etc.) to get the dependence of the properties (e.g., $\omega(V)$). The dependence of the property is then substituted into the model formulas for transition temperatures, conductivities, etc. to calculate the properties that are finally in practical demand

performed by the kernel are combined to construct the evaluation of various quantities such as conductivity or spectrum in the "outer loop" which repeatedly calls single-point calculations.

For example, when you hear that the coefficient of thermal expansion of a solid is calculated by ab initio package, the coefficient of thermal expansion α is NOT "a value directly obtained by solving equations". The thermal expansion coefficient is evaluated using an analytic formula based on Gruneisen's theory [6],

$$\alpha \propto -\frac{\partial (\ln \omega)}{\partial (\ln V)} . \tag{1.2}$$

The ω which appears in the formula is the phonon frequency, which can be obtained as the quadratic curvature of the fitting curve applied to the energy dependence on the atomic displacement as shown in Fig. 1.2a.

By repeating such evaluations of ω with gradually modifying the volume V, then we can further estimate the dependence $\omega(V)$ as shown in Fig. 1.2b. By taking the numerical differentiation required in Eq. (1.2) applied to the $\omega(V)$, the coefficient of thermal expansion could be calculated. Again, it is NOT "a value directly obtained by solving equations (kernel calculation)" but the quantity obtained by "the outer loop", i.e., that composed of the repeated calculations.

To summarize the points again in line with the metaphor of the "**reactor core**", it follows as

- Properties are calculated by examining the *dependence* of "the energy value obtained from the kernel calculation" on the *calculation conditions* such as the atomic positions, volume, etc.
- Ab initio calculations are used it not as a tool to predict absolute values of properties, but as a tool for "trend prediction", i.e., whether the property will increase or decrease as we substitute the elements (Sect. 1.3.2).

For the way to perceive the problems using ab initio methods in this manner, let us call it as Ψ_0-**modeling** in this book. Ψ_0-modeling will be discussed in more detail in the next section. It will also be reiterated in Sect. 1.2.3 as well as in Sect. 6.2.1 in more detail after completing to learn theoretical background.

As will be explained in detail from Sect. 1.3.1, the matters that beginners should take time to learn are the various concepts specific to the handling of the kernel (single-point calculation). These concepts are **the most novel issues for beginners**. Once one has mastered these concepts, other topics associated with the outer loop calculations are not so new things. If you are a practitioner who has worked on various properties (e.g., conductivity, spectra, etc.) and would like to evaluate them by simulation, these topics associated with the outer loop are almost knowledge that you have already learned as basic knowledge in the field of materials science. Once you have mastered the concepts associated with the kernel part, which is the goal of this book, you will be able to understand the specialized jargon that appears in papers treating ab initio analysis. By learning this, you will also be able to judge to what extent you can trust the reliability of the theoretical predictions described in the paper. As such, from onward, this book will focus on the subject of **understanding the concept behind single-point calculation**, which has been metaphorically described as "the core of a reactor".

1.2   What May Be Confusing to Beginners

1.2.1   Not What One Would Normally Expect from a Simulation Study

In the single-point calculation (kernel of ab initio package), what is solved is always the same equation, no matter what bonding form (covalent/ionic/intermolecular...) governs the object, whether it is solid, molecular, metal, or insulators. It is really a *sweet deal* that only one equation can handle everything. There is, however, a price to be paid for such a sweet deal. This is one of the source of difficulty for beginners to understand the framework, as explained in order in this section.

Although ab initio calculation is surely one of the methods in simulation science, the structure of the framework is all inverted from what non-experts might recall from the word of "simulation" (Fig. 1.3). The simulations that are generally recalled by non-experts of science would be the simulation of disaster such as tsunamis, typhoons, and virus pandemics. Other examples include the simulations of fluid for designing aircraft and ships, and the simulations describing population and currency fluctuations. Even for those close to the field of materials science, what are easily recalled as "simulations" would include that of macroscopic strength against vibrations of material in structural mechanics, that of electromagnetic field in electronic device products, and that related to synthesis and reaction processes in chemical engineering [7]. In the case of microscopic level, molecular dynamics simulations would be the first things that come to mind.

The examples listed above are all **phenomenological simulation** and are in a counterpart relationship to the ab initio **simulation** explained in this book (Fig. 1.3). In phenomenological simulation, a **model** is first set up for the phenomenon to be described. Based on the quantitative relationship observed for the phenomenon, the scientist sets up a hypothesis of the laws governing the relationship (the model), on which the computer simulates the quantitative behavior under a given condition. Although I mentioned that the governing laws are given as hypotheses, some of the governing laws may be "firmly established equations", such as the Navier–Stokes equations or Maxwell's equations, that do not fit the wording "hypothesis". However, when solving the equation, there must be parameters to be given to the equation, and the user's subjectivity intervenes as to why those parameter values are set. "The set of equations to be constructed and the associated parameters" correspond to what is called a hypothesis about the laws governing.

In phenomenological simulations, the practitioner's "brain is consumed" in the insight and discussion to justify why that model was applied and why that range of the

	Model & Parameter	Relation between results and Phenomena
Ab initio	Always fixed procedure	Intelligence required
Phenomenological	Intelligence required	Results directly explain phenomena

Fig. 1.3 Phenomenological simulation versus ab initio simulation. The "brain use" part is different for each of them

parameter was selected, which is the *scientific part* in this field. In these simulations, the output variables of the model directly represent the target phenomenon to be described. For example, in the case of electromagnetic field analysis, the output of the simulation is the electromagnetic field distribution of interest, and in the case of population change, the output of the model is the population itself. Getting such an appropriate model is what is obtained for the payment by researcher's brain.

On the other hand, what about ab initio simulations? The equations to be solved are fixed, and, in principle, the only parameters to be given by the user are the atomic number and the position at which the atoms will be placed, those are obvious things.[2] In this sense, there is not much room for consuming "the brain of researcher" to set up the model.[3] However, there is "a price to pay", namely, there is other places where you will have to consume your brain that you did not use much in the model building instead. That is, "bf how the researcher relates the results to the phenomena". In the most stoic terms, the only outputs obtained from the ab initio approach are the ground energy and the charge densities. In DFT method (the topic dealt in this book), we get one more piece of information, the band dispersion, as the "electronic structure", i.e., detailed situation at the electronic level in the system of interest. However, all of this is "**evaluated value at zero temperature**". Remember that the practitioners are usually never doing ab initio simulations because he/she wants to know the band dispersion at $T = 0$, but rather because he/she wants to know the practical properties at finite temperature such as optical absorption spectrum or the thermal conductivity. In the case of the ab initio approach, the results obtained from the simulation do not directly describe the phenomenon to be concerned. The "science of modeling" is required to reconstruct the results of ab initio calculations to **link them to the phenomenon**.

For example, suppose a problem: "Cracking occurs in the metal that is used to attach the device on the substrate, so we want to add an atomic element to the metal to suppress cracking for increased yield of products. There is a theory accounting for the cracks at the interface as caused by the interface being pushed because of too unbalanced ionic transfer velocities of the two elements across the interface [8]. Another theory states that the ionic transfer velocity is determined by the height of the energy barrier when the position of the ion is moved along the path [2]. Here we calculate the energy values for various geometries with different ionic positions along the path to get the estimation of the energy barrier from plotting the energy

---

[2] Nevertheless, in practice, there are several detailed parameters that should be determined and given by the user, and understanding how to set these parameters is the goal of this book. However, these parameters have a slightly different meaning from "parameter selection in phenomenological simulation". You will learn about this feeling from onward at Sect. 4.

[3] The orthodox approach to understanding natural phenomena has long been "model-building science", being opposite to the ab initio approach apparently. This would be the reason why the ab initio research has not been very popular in some countries, especially among basic researchers in the physics, leading to structural difficulties to attract talented people to this field. in the industry in attracting talented people to this field. In some cases, this is a structural difficulty for the industry in attracting talented people to this field.

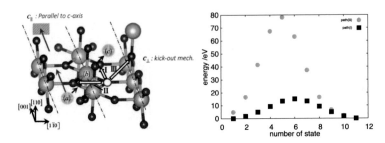

Fig. 1.4 When the ion of interest is moved along the path [white arrows in panel (**a**)], the height of the energy barrier can be evaluated using ab initio calculations [panel (**b**)]. Instead of calculating the ion velocity itself, we can evaluate whether the ionic velocity can be reduced or increased by estimating the increase or decrease in the barrier height [9]

values [9] (Fig. 1.4). If we can capture the trend that the energy barrier increases or decreases with the atomic doping, we can predict that the doping suppresses the ion transfer and hence the cracking. This means that we have achieved our goal just by using the "**black box** that can only evaluate the **ground state energy**". As such in ab initio approach, the "brain" of a practical researcher should be consumed not *inside* the simulation framework but *outside*, namely, "how to combine the results with the target phenomena" by combining the background theories for the phenomena intelligently (Fig. 1.3).

Tea Break; Ab initio approach was hard to be understood

When the author was first appointed to this position, there were no faculty members in the department who were familiar with the new concept of ab initio approach, and I had to deal with many comments and criticism of "off the point", such as "It makes no sense as research to conduct simulations on a subject for which experiments exist". The hardest experience were when we are asked "What are your innovations in model building for your study?" and "What criteria did you use to select the parameters?". Even when I politely explained the difference in approach between phenomenological simulation and ab initio simulation, he would not listen to me at all, perhaps due to his pride on his own career, and would kick me out, repeating "Central importance in simulation research is about model building and parameter selection!". Even my student's research had been accepted by a high-impact international journal, it is evaluated as the lowest score in the PhD examination.

1.2.2   Hierarchical Structure of Scientific Pictures

The properties that can be utilized as useful functionality are usually dynamics at finite temperatures or transition dynamics under the competing balance (e.g., local-ized electrons at the the edge of the instability toward itinerary). Although the root of all these phenomena is, of course, the electronic structure determined at zero temperature, it is a long way to go to deal with such dynamics from ab initio prediction of the electronic structure. Metaphorically speaking, it is a very long way to elucidate a nation's economy from the level of the human genome just because it is humans who form the economic activity. Such an effort is usually regarded as a reckless strategy that does not take hierarchical view into consideration. If we want to study economic activities, we do not need a resolution to analyze the level of individual human beings governed by the genome. Moreover, if we want to discuss economic trends at the national level, it would be appropriate to set up a model in which "industry" or "government" are the minimum units of analysis to represent the averaged quantities over individuals. What forms the economic activity of a nation is surely the economic relations (economic interactions) among individuals, from microscopic viewpoint. However, if we are discussing the macroeconomy at the national level, the usual approach is to take a model on the effective economic interactions between the government, industry, etc. those are the macroscopic degrees of freedom after the averaged over the individuals. This is the standard approach in science that takes hierarchy into account.

For materials properties as well, it is usually treated in the view of effective interactions among the degree of freedom of interest. It is common to deal with free energy model that describes the dynamics of the degree of freedom at finite temperatures (model theory). Simulations for model theory are called **model calculations**. For example, it is the polarized distribution of electric charge that forms polarization in a material, and the charge distribution can surely be evaluated by ab initio calculations. However, it is not usual to perform simulations from ab initio to analyze the dynamics between polarizations. The degrees of freedom of interest are modeled as polarization moments, and the effective interaction between the moments is set up to write down in a free energy model. Then one can apply the framework of the statistical mechanics at finite temperature to perform the model calculation. Model calculations are also phenomenological simulations, where the user needs to set **empirical parameters**.

! Attention

Among the quantities composed from a large number of single-point calculations using the ab initio approach (Fig. 1.2), there are, for example, the thermal conductivity, various optical coefficients (refractive index, etc.). In the ab initio approach, these are "the quantities obtained by calculation (calculation results)", but in phenomenological simulations, these are "quantities set manually by the user (input to calculations)" as empirical parameters, and are used for simulating more advanced dynamics, such as behavior at finite temperatures. In this context, ab initio simulations are referred to as **non-empirical** simulations.

Here, we can conceive the opposing relationship of "ab initio **versus model**", and for a long time, it has been possible to separate whether this is "model calculations" or "ab initio calculations". In recent years, however, ab initio packages have evolved to incorporate even the "outer loop" for model formula part (Fig. 1.2) into the package, and to be able to calculate properties other than energy values, such as thermal conductivity and superconducting transition temperature, as output of the package. These properties are not in a hierarchy that ab initio can handle directly, but are evaluated by the formulas derived by model theories. For these formulas, one should provide parameters like the coupling constants for the electron lattice interactions, etc. as the origin of thermal conduction or superconductivity, which were previously set by empirical choice. These parameters, however, can, in principle, be calculated using ab initio simulations, as the rate of change in the energy when lattice distortion is applied. Contemporary ab initio packages have made up the link to provide the parameters calculated by ab initio calculations into the model formulas to evaluate the properties such as the conductivity or transition temperature inside each package.[4]

1.2.3   Kernel Calculation and Outer Loop

The linkage of providing parameters calculated by ab initio calculations to model formulas can be achieved in the following way. First, "a tool" (**kernel**) is prepared that can calculate an energy for a given geometry of atomic positions, and the kernel is enclosed in an **outer loop** that generates a series of geometries as input files corresponding to a series of changes of atomic positions corresponding to lattice distortion. This loop calls the kernel one after another, calculates the energy values (**outer loop calculation**) to form a set of plots to be fit by a curve (Fig. 1.2) representing the rate of change. By fitting its expanding coefficients, we can estimate the electron–lattice coupling, etc., which can then be substituted into the model formulas to obtain properties such as transition temperatures.

When the author was a postdoctoral researcher, the above sequence of steps was done by human power. Nowadays, kernel calculations can be completed in a few seconds, and automatic processing such as script processing ($\rightarrow$ Sect. 3.4.2) has become widespread enough that procedures such as "generating input files one after another" and "plotting, fitting, and calculating the rate of change" can be automated by programs. Contemporary ab initio packages have come to include that level of functionality. It is then inevitable that a beginner would misunderstand that ab initio calculation is something that gives us thermal conductivity or transition temperature as its outputs. However, ab initio calculations originally refer to the kernel calculation part, so when you start reading a book to learn ab initio calculations, you will find that it only talks about the kernel calculation, and you will be confused, thinking that this is not what you want to learn (Fig. 1.1).

---

[4] When I was a graduate student, I was taught that this was an approach that should never be taken. We were taught that it was outrageous to predict trends by changing the parameters on such model theories that would involve delicate dependencies through exponential functions.

1.3   What Is the Goal to Aim at as a Practitioner?

1.3.1   Why Learn the Kernel Calculation Part?

Then, the question is why should a practitioner interested in the results produced by outer loop bother to learn the concepts of kernel calculations? From a pragmatic standpoint, all a practitioner needs is **the ability to edit the input file**, and the purpose of this book is actually to understand how to do it. There are a variety of parameters in the input file, and if you try to understand all of them in a flat manner, you will end up spending days reading a huge amount of manuals. Some of the parameters, such as "how many times to iterate the outer loop" or "in which directory the input file is located", are things that can be easily understood by reading the manual without asking a master, even if you do not have much technical knowledge.

A teaching course for beginners should select and restrict the topics on the most difficult concepts carefully. These topics are nothing but the input parameters involved in the kernel calculation part. Conversely, once the learning goal has become clear so that one may understand the input parameters involved in the **kernel calculation**, the essence of the background theory for ab initio calculation can be traced naturally. This makes what you thought was a difficult theory much more accessible. The biggest benefit in practice is that you will be able to read the theoretical part of the paper in a smooth manner. When the author was conducting joint research with industrial engineers, the first thing they asked me was this point. "When I read the paper, I see the statement saying 'The k mesh is set to XX, and E_{cut} are set to YY. We used ZZ for the pseudopotential, and AA for the exchange-correlation functionals ...' for which I cannot understand at all. As such, I want to be able to read and understand this part". This is exactly the input parameter for the kernel calculation. As is clear from the fact that these are the contents that should be clearly stated in the paper, the results predicted by the simulation critically depend on the choice of these parameters, being the key to the reliability of the theoretical predictions presented in the paper. Conversely, this is the only difficult part to understand and once you have completed this part, you will be able to use the ab initio calculation package on your own.

1.3.2   Tips for Surveying Preceding Studies

The tip for successful learning also lies in how to choose your first target system. Practitioners usually come to learn the ab initio method with a very specific application in mind, such as the materials tuning of catalytic properties of a surface, etc. As such, they might tend to start with their realistic target of complicated geometry, which is an **absolute bad idea to be avoided**.

The beginning practitioner tends to try to put as much detail as possible into ab initio calculations for the system he or she is interested in. For example, to capture the problem in a surface system with a very complex surface orientation, one would have to take a considerable size in both the in-plane and layer directions. However, such a size cannot be handled in an ab initio calculation, even using a national flagship supercomputer. When I tell them this, beginning practitioners often say, "Then ab initio calculations are completely useless for my problem!" and they may lose all interest in learning.

However, when we do a literature search, we usually find articles reporting that ab initio calculation is used to analyze the catalytic properties of material surfaces. There, "the problem insights" have well been made such as "the essence of the catalytic properties of this material lies in the interatomic binding forces at this atomic site" or "the binding forces near the surface, not in the bulk, dominate the essence of the properties". In this way, the problem is finally formulated in more simplified way to the question of how the binding energy for a single adsorbed molecule varies with the orientation of the surface composed of only three atomic layers. Researcher's "brain" is used to reduce the size and geometry of the system to a minimum that describes the essential factors, and to reduce the size of the problem to a size that can be used in ab initio calculations. If the surface orientation is too complex, we map the problem into a simpler problem that only captures the essence of the problem.

It is sometimes found such practitioners who think "No, it's not possible to make it simplified in my problem", who may be caught in the preconceived notion of ab initio calculations as a tool for evaluating absolute values. This misunderstanding is inevitable to some extent, because one has heard that ab initio calculations realize *quantitatively* precise evaluations in contrast to qualitative model methods. However, the key point here is not to take the ab initio method as a "quantitative tool to be used for absolute value comparison" for the practical materials researches. For the trend of a property (whether it gets larger or smaller when tuned by elemental substitutions, etc.), the method is used to predict it and analyze why the trend occurs where the qualitative predictions are just used to discuss it gets larger or smaller, rather than comparing absolute values with experimental values. For example, when the trend in some experimental values is observed as "103.45 for Ni-substitution, 106.32 for Co-, and 208.72 for Mg-substitution", **we cannot simulate the whole in its entirety**, so we reduce it to a simplified problem as described above. Applying the ab initio simulation to the simplified model problem, then we would get the trend "501.36 for Ni-, 505.77 for Co-, and 604.34 for Mg-substitution" which can be used as a justification of the observed trend. You may be surprised to hear that 103 is reproduced as 501. If they are energy values, you have to remember that the absolute value of energy can be very different depending on how the reference zero is taken. In ab initio calculations, only the energy difference is meaningful.[5] If the calculated values are not energy values but elastic constants or thermal conductivities, etc., they should be roughly consistent with the measured values in terms of orders and ranges, but if you remember that the calculation is made for a simplified model with computable size being not a full

---

[5] It is called ΔSCF (delta SCF) in the sense of difference. For the meaning of the abbreviation "SCF", see Sect. 3.1.

simulation, you will understand that it is not appropriate to use the estimation to expect a precise agreement in absolute values observed in experiments.

> ## ! Attention
>
> Although we explained the contrast between ab initio and model calculations in the previous section, it seems ended up saying, "Even ab initio results are modeled one". In the end, modeling is necessary in anyway to understand the mechanism of a phenomenon, but the contrast is whether it is done *inside* or *outside* the simulation. In ab initio analysis, the tool itself is not modeled, but for the target object to which it is applied ("outside" of the tool) it is needs to be modeled, as explained in this section. Before practical applications blossomed until around year 2000, most of the research on ab initio calculations involved methodological development. Since these were problems of improving the quantitative descriptiveness of prototypes such as silicon solids and hydrogen molecules, the research aspect of "understanding the phenomena by modeling the subject" did not come up much. This may be the reason why the contrast of the approaches between "model theory" versus "ab initio" was sharply distinguished.

In the examples explained here, a two-step modeling scheme should be made at the research planning stage: first, how to describe the catalytic properties of the surface by ab initio calculation (the tool just provides ground state energy) as **"1/Nano-modeling strategy"**, and second how to cut down the typical aspects of the target system into computable sizes as **"2/Modeling of the target geometry"**.

One may be wondering how a novice practitioner could possibly accomplish such a modeling of the problem which seems to require much intelligence. The answer, however, is surprisingly simple. If one searches for "surface catalyst DFT" on Internet, one will find similar preceding studies, so you can just read through these papers, asking "How is Nano-modeling done? How is the modeling of the down-sized geometry done?" to get the ideas of the modeling in the paper.[6]

1.3.3 Finding a Tracing Calculation Task

Once you have found a similar preceding study, the next thing to do is to **trace the same calculation** with that similar preceding study. It goes without saying that in quantitative research, one should proceed carefully with proper calibration. You should never start with your final target system from scratch, because even if you get a predictive value, you will not be able to judge whether it is a **reasonable value or not**. As you will learn in detail later in Sect. 4, the prediction can *vary greatly* depending on computational conditions.

---

[6] It is assumed that there are similar preceding studies. If this is not the case, establishing such modeling without any preceding works by one own would be a first-class achievement.

First, you need to start by reproducing the values that have been reported as reliable achievements in papers, and then proceed in the form of tapping on a stone bridge while taking calibration and checking how much the results have changed as a result of what conditions you have changed. After first ensuring that the values reported in the paper can be reproduced properly, you can gradually move on to your final desired problem by moving the input settings around. A proper paper should describe the minimum information about computational conditions so that the reported values can be properly reproduced. And those conditions are the **specifications for kernel calculations** that you will learn about in this book.

1.3.4   Learning Approach to Be Avoided

Having explained so far, the readers were expected to wonder that there is the learning approach that should definitely avoid by the practitioners at the beginning level. That is, one should NOT start with the books titled such as "Fundamentals of Density Functional Theory". These are not practical books, but books that explain the DFT as a fundamental theory of many-body electron theory, which answer to the purely basic scientific question, "Why can we get to an exact solution of the interacting many-body problem by solving the one-body form?" [5]. The answer to this question was given in the 1960s, mainly by U.S. Nobel laureates in Chemistry, and a series of these achievements are labeled as "Density Functional Theory". At its core, it is a mathematical proof using the "proof by contradiction", and many abstract and difficult problem settings appear, e.g., "v-representativity", etc.

If one is aiming at being a theoretician majoring many-body electron theory, one should have read up on these mathematical proofs and the historical background that led to their conception, but practitioners do not need to read through these mathematical proofs in advance in order to start using DFT package. It is all well and good to get started this kind of fundamental topics with intellectual curiosity, but it is completely nonsense if one becomes disheartened by the abstract and esoteric content and lose the desire to learn. A recommended way is that first complete the general overview of the method and a rough understanding of the exchange–correlation functional, in particular, from practical books like this book, and then start reading the fundamental books as "the next book to read". By this way, one will be able to read through the fundamental books without stalling out.

! Attention

One of my most often citing stories to explain this kind of instruction is about the books like "Foundations of Quantum Mechanics" such as that by von Neumann [10]. Despite the fact that these books are bulky like two volumes, it does not cover any of the standard topics such as the handling of hydrogen atoms, perturbation theory, variational calculations, etc. The subject of these fundamental books is the problem of mathematically proving that "the space for the general solution of a partial differential equation even with singular divergence term (Coulomb potentials) can be expressed in terms of the expansion by eigenfunctions", explained over two-volume contents. It is a fun to read for those who have once learned over the topic and want to go back and think "Why was this in the first place?", but would be not appropriate for a beginner who wants to know what quantum mechanics is all about. This is an example of why it is not the fundamental books for beginners to start with.

1.3.5   Goal to Aim at for Beginners

So far, we have explained the need to understand "the specifications for kernel calculations" and the tips of starting with trace calculations. When starting to learn this method as a practitioner, it is important to keep in mind the goal you are trying to achieve, and to be aware of how far you have come, as shown in Fig. 1.5.

[1st-grade] *Goal; Able to trace the exercises provided in a course.*

> **(1a) Able to understand the outline of the procedure using example exercises provided in each package, to understand the terminology appearing in the procedure.*
> **(1b) Able to understand the meaning of each parameters in an input file at elementary level, to adjust them as one wants.*
> **(1c) Able to understand theoretical concepts such as SCF convergence, XC potentials, and pseudopotentials.*

[2nd-grade] *Goal; Able to trace the preceding publications.*
> *Able to perform calculations at a publishable level under a supervisor.*

> **(2a) Able to generate input (molecular/crystal) structure on their own*
> **(2b) Able to understand the specs presented in preceding papers and reproduce the results.*
> **(2c) Able to produce necessary figures.*
> **(2d) Even if a calculation gets stuck, able to solve the trouble by adjusting the computational specs. [e.g./Choice of smearing, convergence algorithm etc.]*

[3rd-grade] *Goal; Able to formulate a similar project on their own by understanding similar preceding studies, to initiate the project as a concrete calculation procedures.*

> **(3a) Can teach others how to use the one's "primary" software (e.g., CASTEP/GAUSSIAN) in which one has the most experience.*
> **(3b) Able to understand and explain comparisons and contrasts between one's primary software and others in compatible method (e.g., CASTEP and VASP).*
> **(3c) Able to understand and explain comparisons and contrasts between one's primary and others with comparable method (e.g., DFT and MO).*

[4th-grade] *Goal; Milestones to be targeted for further study.*

> **(4a) Able to understand and explain a wide variety of electronic structure methods and what solutions they offer (e.g., GW, DMC, BSE).*
> **(4b) Able to understand which package provides what kind of solution, and be able to combine the appropriate package for a given target practical problem and incorporate it into an analysis using electronic structure simulation (modeling of the problem).*

> **Goals and checklist items for beginner practitioners.**
> **This book aims to master the contents of the 1st-grade.**

Fig. 1.5 Achievement level that the beginning practitioner should be looking for. This book aims to master the contents of Level 1

1.4   Why It Is an Opportune Time to Jump in

For those who have picked up this book, it is probably known that the ab initio simulation is at the good timing for researchers in materials science to get. But why is it an opportunity? In this section, we explain about this.

1.4.1   Affinity with Data Science

By definition, the research using model calculations takes the "tailor-made" style for each applied object. In contrast, **the great strength** of the ab initio approach lies in the fact that it can handle everything in only one format and in only one package (only one kernel), whether it is an insulator formed by covalent bonds or a biomolecular system formed by intermolecular bonds. Since variations in input files alone can cover a wide variety of substances, it is possible to ensure a great deal of versatility by devising an outer loop that generates input files to be fed into the kernel. As described in Sect. 1.2.3 about generating input files by an outer loop, ab initio researches have significantly transformed as bf Material Informatics in conjunction with data science, machine learning, and artificial intelligence techniques.

> **Tea Break; Understand the amazement feeling in ab initio approach**
>
> Since the author himself came from the field of model analysis as an outsider to the ab initio approach, it was quite difficult to switch my mindset from the model approach. What would be interesting if we gave an atomic geometry of silicon solids and solved the equations and got the insulator correctly? One thing to realize is that this is NOT "build a model of an insulator and an insulator comes out" (that's surely obvious). As one great professor carefully explained to me, "Just a single equation really brings out the diversity of matter. It may be true in theory, but when you calculate it with a computer, it is an honest surprise to see the diversity of insulators and metals really appear in front of your eyes as numerical values". It may be a matter of course for the generation that has already been familiar with the package as already there, but the impression of the generation that constructed this framework was "I didn't think it would really come out properly", they said. They were not expecting it to really appear after they had simplified the problem for numerical calculations, such as the adiabatic approximation and non-relativistic approximation.

1.4.2   Outer Loop Linked to Artificial Intelligence

The ab initio electronic structure calculation itself has been well "matured", and the convenient platforms such as packages had been available for a long time from 1980s. However, the situation has changed drastically around the 2010s in the author's impression. Until then, the main research interest was the methodological development by basic researchers of **many-body electronic theory**. But the ab initio calculations have undergone a qualitative change since then.

This "significant change" is due to a deeper background than simply the fact that the range of practical applications has increased because of the increased computing power. If DFT was originally developed as a framework for simulating electronic properties of materials on computers in a practical manner, there would be no change in the direction that experts in the field are aiming for, and it would not be a *qualitative* change, merely an acceleration of progress in the field. Rather, it is important to note that the *mindset* of the field has changed. In fact, rather than being a numerical method for practical use, DFT should be more appropriately viewed as an important and fundamental theoretical framework related to many-body electron theory, which was "discovered" in the 1960s. The positioning of the theory is that it provided an affirmative answer to the purely academic question, "Can an interacting electron system be projected **strictly** onto a 'solvable one-body problem'?" [5]

Tea Break; DFT appears in lecture course?

Often this is not well understood in some countries, and then DFT has been excluded from the required topics in university lectures, even that is quite fundamental topic in Physics. There is a guideline that the numerous experimental techniques are not to be included in the lecture course but rather should be given in each individual laboratory. DFT has actually been misunderstood as "a numerical simulation method" and is considered to be on the same level as individual experimental techniques. As such, even in "majors related to many-body electron systems (whether experimental or theoretical)", there was a situation in which students got PhD without even knowing the existence of the problem setting of "whether it is possible to project interacting electron systems into a strictly 'solvable one-body problem'" with affirmative answer.

Since DFT was developed with the above positioning, the main focus of research was originally not on practical application to real materials, but on *verification of the theory*. The method was applied to real materials only as a benchmark for demonstration, and research was conducted on simple systems such as silicon solids and simple metals.

An expert in many-body electron theory could stop at the level with most stoic proverb that the *physical* quantity obtained by DFT is only the ground energy and the charge density, while the band structure is merely an intermediate quantity for mathematical convenience which is never ensured to have physical meaning. Though the calculation used to be very costly where a single-point calculation takes a day on large supercomputers, it gets faster to be possible to completed on laboratory servers and even on laptop PC nowadays. Then the increasing number of practitioners start to utilize the band structure results for their materials researches for making interpretations on their experimental results. Some researchers in the field began to realize that "this tool can be used not for predicting absolute values, but for predicting the qualitative trends of properties when atomic structure tuning is made in materials properties". This is how the use of ab initio calculations became widespread in experimental analysis and planning in the 2000s.

Tea Break; Application to analyzing experiments

An old professor in this field recalled that when he went abroad to study this method in 1970s, he was told by seniors that it is not a proper scientific manner to make quantitative comparison between DFT and experimental values arguing like "$xx\%$ in agreement with the observed value". What they insisted is that even though it is ab initio, it boldly approximates reality using the non-relativistic and adiabatic approximations, and hence it would be no sense to directly compare the results of such calculations with experiments. However, when he arrived at host group abroad, he was really surprised to find that everyone was making such argument to compare the DFT results with experiments that was taught not to be made when he departed.

In this book, the electronic structure calculation for a single input file to define a geometry is referred to as a "single-point calculation" (Fig. 1.2). Due to improvements in computing power, these single-point calculations can now be performed in a few seconds. This has led to the emergence of studies that rapidly generate a series of geometries for the input file, calculate their energies to be checked if lowered or not in order to find novel structures that give lower energies (**structure search**). The original idea was to use ab initio calculations with computational resources to discover novel unknown crystal structures realized under extreme conditions such as in the mantle of the earth's crust or in outer space.

The initial strategy for structural search was to randomly generate geometry (**random search**). Since this was a time when data science and artificial intelligence research was getting successful, the random search is considered to be improved for more efficiency to get successful structure assisted by the technique such as genetic algorithms or Bayesian sampling. This formed one of the trends toward **Materials Informatics**, which is the fusion of **artificial intelligence and machine learning with Materials science** Chap. 7. A similar development is the research on autonomous systems in which artificial intelligence is applied to the *outer loop* level to control the implementation of mechatronics, which is a fusion of mechanical and electronic engineering. In our topic, artificial intelligence is being applied to the *outer loop* to call ab initio single-point calculation to form a new field, Materials informatics.

1.4.3 Confluence with Computational Thermodynamics

In the 2010s, the method merged significantly with computational thermodynamics, and it seems that the method has largely transformed itself into a tool for materials development. One may be thinking, "Isn't ab initio calculation originally a tool for materials development?" However, as I have mentioned several times, stoically speaking, ab initio calculations can only describe the ground state at zero temperature. Most of the useful materials properties involve dynamics such as excitation and transport at finite temperature.

Such materials properties are governed by the competing tendencies between "that toward lower energy in the ground state (enthalpy effect)" and "that toward getting high number of possible configurations to be accessible (entropy effect)".

One of the expected roles of theoretical materials science is to analyze the **free energy model** for this competing situation. If enthalpy and entropy effects can be mathematically modeled and given, the predication framework is well completed, and a framework for processing this by computer has been established in the field of computational thermodynamics [11,12]. Using this framework, it is possible to predict various materials properties, such as the phase diagram of steel and the calculation of melting points.

The parameters required for the free energy model have been provided by a set of numerical data obtained from the measurement of lattice specific heat (corresponding to the degrees of freedom of lattice vibration) in the past. However, as ab initio calculations get faster and faster, it has become possible to perform the evaluation of the specific heat as a variations of the "outer loop". Namely, by applying distortions corresponding to the normal modes of a crystal, we can get a set of phonon frequencies as the curvature of the energy change. Once the set of frequencies are given, we can estimate the specific heat by using several model formulae like Debye model, etc. Obtaining accurate data through experiments is time-consuming and costly, and in some cases, such as material phases under extreme conditions, the experiments themselves are difficult, which has been a limiting factor in the computational thermodynamics. Nowadays, ab initio simulations can supplement such data.

Among the entropy contributions in the free energy model, the lattice specific heat (corresponding to the degrees of freedom of lattice vibrations) can be evaluated by ab initio calculations as above. In addition to that, various other entropy factors are involved in the free energy model, such as the degrees of freedom of atomic vacancies and impurities, and the degrees of freedom of electronic spins. Theoretical simulation fields to describe each of these factors have been established [12]. Ab initio calculations are now fused with these fields and positioned as a tool that contributes to the construction of the **free energy model** (providing a part of enthalpy and lattice vibration terms), which has been integrated into the field of computational thermodynamics and transformed into a form that plays a role in "theory for materials development".

1.4.4 Materials Genome

Although the DFT initially developed as a purely theoretical framework to answer the fundamental question of "Can the interacting electron systems be reduced to the feasible one-body form?", it have been linked to the exit as a tool for materials exploration as described above. This has given rise to a movement to convert the power of computation and the required electric power into the accumulation of human knowledge for exploring novel useful materials. In conjunction with computational thermodynamics, there is also a growing awareness of what parameters should be provided more by ab initio calculations. For example, if there are parameters such as "energy cost/gain when interfaces or impurities are introduced in a material (formation energy)", "temperature dependence of entropy and specific heat expanded as the power of T", etc., then computational thermodynamics can take these parameters and link them to more practical predictions of material properties not within the extent of ground state energy at zero temperature.

If these required parameters are calculated for all possible materials and a database is constructed, they will become the knowledge of mankind regarding materials exploration. The question is how to exhaust all possible materials. All possible materials are the subset of all possible combinations of at most 100 elements in the periodic table. Since the number of combinations grows as a power of the number of elements, it becomes too large to count even for "at most 100 elements". This is where data science contributes. The materials, i.e., patterns of elemental combinations, existing in the world are now well databased and maintained in a form that is easy to feed to artificial intelligence. By the machine learning over the database, it would be possible for artificial intelligence to propose such combinations that are not confirmed to exist yet, but are likely to exist as compounds, or combinations that are not so likely to exist. Such a framework has already been put to practical use in various ways (for example, predicting "crimes are likely to occur in this area today" based on a thorough examination of data on past crime locations and days of the week).

The concept of accumulating knowledge through a series of automated processes is gradually being implemented: generating possible elemental combination structures one after another, running ab initio calculations one after another on a national flagship supercomputer, automatically processing the results, and registering them in a database. It is called the "Material Genome" Project [13] in analogy to the Genome Project, which uses computers to decipher the genetic information (genome) of humans and other species and leave it as human knowledge for future.

1.4.5   Chance for Latecomers

Even if the field has potential, it is difficult to find any place to play for those who join later after the community has already been dominated by their predecessors for generations. This is especially true for practitioners that have already completed their student years. In fact, communities of ab initio calculation in each country have already got matured with several leading "families" with "big boss". In some countries with strong background for fundamental science, the community has been producing researchers who have been part of the international core since the early days in the 1960s. They are the proud communities that understand the method not as a "black box tool" but from its fundamental theory and have made first-class contributions to the methodology developments.

However, the brewed pride over several generations often has a disadvantageous aspect as well. In some countries, it is hard to deny the tendency to place methodological researches above application researches for use as a tool. The author has often heard from industrial researchers seeking the collaboration with academia that they tried to contact renowned group in ab initio community (before visiting the author), but they were met with a salty response, saying, "We are not really interested in such a practical problems".

The author can somehow understand this feeling to some extent, from personal experience as grown up as a postgraduate student in such field with proud for several generations achieving world-leading contribution in the past. After the early days of the field when it was just starting up, many research themes get to be subdivided into XX-category, YY-category, etc. Then the later generation who grows up within the already established community becomes aware that XX is the classic" than other (so-called "mounting" consciousness).

It is worth noting that the consciousness on "which is the classic" can easily be reversed when across countries and communities. As the thermodynamics is known as a typical example, the structure is not always such that "fundamental theory are the first and applications come downstream". In some cultures, even a background in calculus or linear algebra is considered a mere skill, on the same level as being able to use a screw driver. Others believe that the science dealt with in higher education, regardless of whether it is in the humanities or sciences, should be the intellectual activity of how to put practical problems into schematic models to be analyzed. From this perspective, there is a certain fraction of theoretician in academia who believe that modeling like "Ψ_0-modeling of industrial problems" is an object of intellectual interest.

The fact that the established players are not so interested in practical applications is an advantageous opportunity for newcomers. Once the methodological pioneering has settled down to an acceptable level for practical use, the next tide is turning toward Psi_0-modeling: "How to model the problem into the one for which ab initio approach is applicable?" The more prestigious the group to which one belongs, the more difficult it is for "citizens grown up in the empire" to get out of "classic" mindset. Conversely, active researchers at the forefront are emerging one after another from communities that are different from those of the older school who has majored in methodology from earlier times. For the newer layers of researchers from practical materials research communities such as industries or experimentalists, they would have been ridiculed in the past to be said "They are using a black box without knowing the theory behind it". However, now that the methods have matured sufficiently, we are no longer at such poor level of using them in black-box-wise even if we are not from the classic community. Unlike "constructing a theory", "understanding a theory that has already constructed" is much easier to catch up on.

I used the metaphor of "citizens growing up in an empire". It is instructive to feel the breath of the first generation that built that empire from scratch. The first generation had constructed the community with paying respect and concern for "outside the walls" of neighboring fields, and been aware of their positioning among the other communities in panoramic viewpoint. As I mentioned above, the role of electronic structure calculation is positioned within the broad framework of the free energy model, but in books written around the 1960s, such a way of positioning is discussed as a matter of course. As time goes by and the methodologies in the field become more sophisticated and deepened, the walls become higher and higher, and the overarching positioning of the field is gradually forgotten, coupled with tradition and pride.

Research using ab initio calculations is now in a period of turning the tide, opening up a "rich fishing ground" where the old-timers are unable to change course toward it because of their aversion to muddy practical applications. As for the fundamental theoretical background, let's proceed by first comprehending the whole picture and securing a sense of security by avoiding the "mazes you should avoid entering to get lost" and by grasping the tips, with the help of this book as a guide.

Ψ_0-modeling requires a certain amount of talent to come up with it from scratch, but that is not necessary, don't worry about that. Just searching similar preceding studies to get a sense of "Oh, so this is how to model it", then you can take a hint from them and can apply the same strategy to your own target. You might be surprised to find that the idea of modeling in the preceding studies is so simple and naive. Beginners usually tend to image that the experts are taking so precise approach with firm theoretical formalism that outsiders are difficult to understand it, but for the modeling to capture the materials properties is sometimes "just comparing the stabilization energies", etc., being so simple and anticlimax than they expected. These simple ideas are just like "We can understand it if somebody tells it, but difficult to come up with if nobody tells it".

We mentioned that "the outer loop" is connected to AI techniques (Sect. 1.4.2). Hearing the term "AI", one might think of a firm theoretical framework like that for Physics, but for a user it is not an appropriate way to learn it from the most fundamental level of methodologies in AI (such as how to improve the optimization efficiency for neural network). The best way would be to start with finding preceding works of the AI application to your closest target to get the idea "how it is used available in which package, which properties can be captured by it, what is the limitation...", that is, how one can quickly arrive at the frontier in Material Informatics.

1.5 Contents of the Following Chapters

1.5.1 Policy on Content

In line with what has been mentioned so far, the basic policy of this book is set to carefully explain kernel calculations as the subject matter. For the tutorial on kernel calculations, we take the calculations of periodic crystal systems using planewave basis function implemented in a free DFT package, "Quantum Espresso" [14].[7] Although some readers may be interested in [Gaussian basis/isolated molecule systems] rather than [planewave basis/periodic solid systems],[8] most of the topics in this book explain common concepts for both subjects. It is never a waste of time for practitioners working mainly on the molecular systems to master the rudiments of planewave basis/solid periodic systems, especially when dealing with practical problems such as "molecules placed on a surface". Furthermore, there has been an increase in the number of studies recently that treat isolated molecular systems using planewave basis. We hope such practitioners taking the opportunity to understand the contrast between Gaussian/molecular and planewave/periodic calculations in a unified manner.

For the intended audience, it is assumed that the only skill required is touch-typing. The basic knowledge of Linux operations required in this book is covered in a self-contained appendix (Chap. 9). Readers who are familiar with Linux may skip the corresponding sections giving introductory explanations on it as appropriate.

1.5.2 Teaching with Command Line

Most of the modern packages of electronic state calculation available to practitioners are Graphical User Interface (GUI) environments provided by commercial vendors. When an application is launched, a window opens in which buttons and numerical value choices can be operated with your mouse.

For the packages, a research group in academics from universities or national research institutes conducts research and development of the core part to add new functions for the next version up. On the other hand, for the parts that have matured and been fully tested, commercialized organizations sell and support them as convenient packaged products. The simulation program itself is developed as **source code** [15] written in programming languages such as Fortran90, C, etc., but it is becoming rare that products can be purchased at the source code level.

The source code is converted by a compiler into an executable file (binary)[9] suitable for each processor architecture on each OS (Win/Mac/Linux) [Fig. 1.6]. The package is usually distributed as a product in this binary format. Thus, the purchaser cannot develop extensions for further functionalities and cannot install it to another machines with different environments (OS/architecture), those are possible only at source code level. The reason it is called a "package" is that it is provided as a set that includes not only the main program that performs the electronic structure calculation itself but also accompanying programs (called **utilities**) that perform

---

[7] A free package developed by a group at SISSA (International Graduate School for Advanced Studies) in Italy. It is a professional-use software used in front-line research, with a large users forming a forum where users can ask questions and share their know-how.

[8] The author's research group treats both of these as the coverage of the study.

[9] The files are called "binary" because they are written in binary format as "machine language".

[GUI operating environment (fixed built-in by vendors)] or
[Automation techniques using script works (flexible development by the user)]

Distribution in executable format (binary).   ↑   *Lower stream*

Development at source-code level.   *Upper stream*

Fig. 1.6 Relationship between source code, binaries, and GUI usage. The source code is converted by a compiler into an executable file (binary as "machine language") suitable for each processor architecture on each OS (Win/Mac/Linux). At the binary level, they are machine dependent, and the source code as the "blueprint" of the function is not visible for purchasers. There are two ways to use binaries: convenient usage for beginners using the GUI, and usage using the command line (handling directly with commands while being aware of the existence of binary files)

various postprocessing such as visualization of charge densities, interface utilities so that the output files of a package are further used as input for another package, etc.

Theoretical researchers prefer packages distributed at the source code level. In this case, it is assumed that a user must compile the code all by oneself suitable for one's environment, and hence it gets more difficult to install it. In the case of commercial code, purchase at the source code level is about an order of magnitude more expensive because it allows a user to tune it for extensions and also because it is that developers to disclose their "inside secret" to the public. Theoretical researchers who are part of a team developing a package often share "beta versions" of such source code within their team and use it at the source code level while developing extensions.

To be honest, when discussing with industrial practitioners in collaborations, researchers majoring fundamental methodologies are sometimes very surprised to hear that the convenience and functionality of GUI packages have advanced so much than expected. Since researchers mainly engaged in methodology are usually working at the source code level, they are not following the evolution of usability at the application side as much as the user practitioners. This is one of the aspects where there is a good chance for new entry of beginning practitioners (Sect. 1.4).

To run a package distributed as a binary (suppose its file name is "maezonoX", for instance), you execute

```
% mpirun -np 8 maezonoX
```

on the Linux command line, which is a command to run "maezonoX" with MPI parallel with 8 cores. However, this is a difficult task for beginners because the user has to remember commands such as "mpirun" and options such as "-np 8" one by one. Therefore, the GUI environment displays a button on the screen that corresponds to the execution of "mpirun -np 8 maezonoX", and when the user clicks this button, the system will type and execute the above commands inside the server instead. The user can set options such as "-np 8" in advance on the window opened by "Menu > Preferences". After that, when the user clicks on the "Execute" button, the specified options are read, and the application automatically generates and executes the command with these options added.

This book does not go into the level of the source code developments which is below the "binary use" level, but rather provides a tutorial at the "command line level" to run binaries which is below the GUI usage level (Fig. 1.6). The tutorial at the GUI usage level is actually equivalent to a "user's guide for each program package" and is not suitable for understanding concepts that are common regardless of each **package**. If you get to be familiar with the command line level, you can easily return back to GUI usage being capable to be aware that "I'm pressing the button corresponding to this".

This book is intended for practitioners, i.e., the readers who will eventually become proficient and engage in full-scale simulation practice. Unlike tutorial-level calculations, full-scale calculations will require the use of high-performance cluster servers and supercomputers. In principle, the use of packages on these computers is done on the command line. If you understand the package operation mechanism at the command line level, it is easy to move to this phase, but if you have only GUI usage experience, you will face great difficulties because you do not understand Linux and do not have an image how the command works.

Another important reason for not using a GUI implementation in this tutorial is that it limits the degree of freedom in setting input parameters. GUI implementations provide "recommended input settings" in the form of the choice among "SuperFine/Fine/Normal" in a built-in combination. Fine-tuning of individual parameters is not intended, and customization is difficult.[10] Since the purpose of this book is to learn the outlines of the theory by experiencing how input parameter settings affect the results, the use of "recommended built-in combination" in the GUI implementation is not suitable for this purpose.[11]

As important developmental subjects, we will explain the concept of "high throughput" and "workflow", later in Sect. 6.3. As explained in Sect. 1.4.1, the most powerful attraction of ab initio calculation is its combination with data science. For this advantage of this linkage, if users have to click on each calculation operation with a mouse, neither speeding up nor automating the process can be expected. This advantage is accomplished by the automation techniques realized by script works on the command-line usage.

This is the main reason why this book is designed to be understandable from the command line to ensure a foundation that will lead to large-scale calculations at the supercomputer level, high throughput, and scalability for workflows.

1.5.3   Machine Dependence of the Tutorial Environment

When ab initio calculations are explained in a tutorial, the problem is the variation in the computer environment that the reader must be prepared for. In this sense, the

---

[10] It is something similar to insurance package products for overseas travel.

[11] A tutorial of GUI implementation will become just an operation manual for beginners, which is not very useful from an educational point of view.

environmental dependence is greatly mitigated by teaching at the command line level
rather than at the GUI level.[12] Although the PC usage environment of readers tends to
differ from country to country, the standard assumption would be WinOS, LinuxOS,
or MacOS. Linux servers, of course, but Macintosh machines are based on UNIX,
so they are equipped with a command line inherently. Even for Win machines, it is
not so difficult to prepare a command line environment because WinOS is designed
to make things that originally run on the command line look like they do not.

Nevertheless, even to perform simple tutorial exercises, there is severe complica-
tion of setting up the *environment*: "Does the visualization software work properly?",
"Does the editor work?", "Are the various libraries required for the ab initio package
downloaded and ready to use?". The trouble lies in the fact that the way to set up the
above environmental issues is different for each MacOS, WinOS, etc.

One possible way to solve the machine-dependent problem is to use "virtual
environment" technology. This is a technology that allows a user to use a "virtual"
Linux server as if it existed in the workspace of a Mac or Win machine realized by a
software. However, at least at the time of composing this book, we have experienced
the following problems as obstacles:

1. Procedures to prepare a virtual environment itself become quite cumbersome for
   beginners (see column).
2. Visualization applications cannot open their pop-up windows from a virtual environment (pop-up is possible only from primary OS, not from "the virtual OS
   inside the primary OS"). As such data must be sent back and forth between the
   virtual machine/OS and the real machine/OS.

(1) is not a problem when the tutorial course is taught in a pre-prepared terminal
environment such as a computer-education center. However, if the reader is to do
the course by oneself on one's own PC as is the case with this book, this is a serious
obstacle that makes it difficult to read through.

Based on these observations, this book does not use a virtual environment, but
uses a RaspberryPi, which can be a very inexpensive Linux server.[13] In the next
chapter, we will explain how to install the Linux environment, "Ubuntu OS", on
your RaspberryPi, and you can proceed with the tutorials by connecting a keyboard
and display to the RaspberryPi and operating it directly. If you have a familiar PC
(Win/Mac) at hand, you can use it as a terminal to remotely login to the RaspberryPi
and proceed with the simulation. This remote use of RaspberryPi from your familiar
PC can be regarded as "purchasing a Linux simulation environment as 'expansion
card' for your own PC". At least at the time of writing, this is a much less time-
consuming way for a beginner to build an environment on his/her own PC.

---

[12] Otherwise, the textbook would be full of snapshots of different GUI screens for each environment.
[13] The most original version of this book (written in Japanese) was written using WinPC and MacPC
instead. These will be published separately in due course.

Tea Break; Tutorial performed on virtual environment

Virtual environment such as "VirtualBox" [16] is available for virtualizing Linux on Mac/Win machines. However, they are not suitable for simulation tutorials because it is too slow to run any simulating calculations. As one possible solution to this problem, "Docker" application [17] is known to realize the environment. This tutorial course was initially provided using Docker on a Mac. The instructor has prepared a set of environment settings for using Quantum Espresso, and then compiling it up as an image, which is distributed to the participants. The participants can get the same environment to run Quantum Espresso (called a "container") by launching the distributed image on Docker application, as long as they are able to prepare the Docker on each computer.

Upon the success with Docker/Mac, we had planned the same for Docker/Win in line with Docker's true purpose of being able to share the same workspace regardless of the OS. However, at the time of writing this book, Docker was sometimes unstable for the contents of this tutorials on Win machines. Furthermore, the problem described in (1) was particularly noticeable on Win, making the Docker installation procedure too tricky. Beginners may lose sight of the main stream in tutorials, disturbed by a large amount of pages to explain how to install "Docker". To avoid this, we decided not to adopt Docker.

Although we did not adopt Docker finally, we have found another educational advantage through our experiences to use it in several international tutorial workshops. That is about "the concern (2)" appeared in the main text. This drawback conversely becomes advantage when the beginners overcome it, as we found. In the case of Docker use, due to the problem (2), there are frequent opportunities to make the participants aware that "this is an operation on a local machine" and "this is an operation on a virtual machine". This served as a good exercise to improve their "spatial awareness". The most confusing aspect of command line operation for beginners is the "spatial awareness" of where they are operating at the moment. The best prescription for it is to simply repeat similar operations over and over again.

References

1. "QMC and phonon study of super-hard cubic boron carbon nitride", M.O. Atambo, N.W. Makau, G.O. Amolo, and R. Maezono, Mater. Res. Express 2, 105902 (2015). https://doi.org/10.1088/2053-1591/2/10/105902
2. "Density Functional Theory: A Practical Introduction", David S. Sholl, Janice A. Steckel, Wiley-Interscience (2009/4/13), ISBN-13:978-0470373170
3. "Drug Design Strategies: Computational Techniques and Applications", Lee Banting, Tim Clark (editors), https://doi.org/10.1039/9781849733403
4. "Electronic Structure: Basic Theory and Practical Methods", Richard M. Martin, Cambridge University Press (2004/4/8), ISBN-13:978-0521782852

5. "Density-functional Theory of Atoms And Molecules" (International Series of Monographs on Chemistry), Robert G. Parr, Yang Weitao, Oxford University Press USA (1989/1/1), ISBN-13:978-0195092769

6. "Principles of the Theory of Solids: Second Edition", J.M. Ziman, Cambridge University Press (1979/11/29), ISBN-13:978-0521297332

7. "An Introduction to Computer Simulation Methods: Applications to Physical Systems", Harvey Gould, Jan Tobochnik, Amazon Digital Services (2017/9/18), ISBN-13:978-1974427475

8. "Diffusion in Solids: Fundamentals, Methods, Materials, Diffusion-Controlled Processes" (Springer Series in Solid-State Sciences, 155) Helmut Mehrer, Springer (2007/7/9), ISBN-13:978-3540714866

9. "Ti interstitial flows giving rutile TiO_2 reoxidation process enhanced in (001) surface", T. Ichibha, A. Benali, K. Hongo, and R. Maezono, Phys. Rev. Mater. 3, 125801 (2019), https://doi.org/10.1103/PhysRevMaterials.3.125801

10. J. von Neumann, Die mathematische Groundlagen der Quantenmechanik, Springer (1932)

11. "Computational Thermodynamics: The Calphad Method", Hans Lukas, Suzana G. Fries, Bo Sundman, Cambridge University Press (2007/7/12), ISBN-13:978-0521868112

12. "CALPHAD (Calculation of Phase Diagrams): A Comprehensive Guide", N. Saunders, A.P. Miodownik, Pergamon (1998/6/23), ISBN-13:978-0080421292

13. https://www.mgi.gov (URL confirmed on 2022.11)

14. https://www.quantum-espresso.org (URL confirmed on 2022.11)

15. "Computational Physics: Problem Solving with Computers", Rubin H. Landau, Manuel J. Paez, Wiley-VCH (1997/8/11), ISBN-13:978-0471115908

16. https://www.virtualbox.org (URL confirmed on 2022.11)

17. https://www.docker.com (URL confirmed on 2022.11)

Preparing Tutorial Environments

2

Abstract

This chapter provides the procedure to build the tutorial environment on his/her own. We construct the environment to run the density functional package "QuantumEspresso". First, we will learn how to assemble the RaspberryPi and how to prepare the command line operation system (terminal environment) on that. Then, we will install various software and utilities. Next, we will prepare the "working place" for the tutorial as a "directory" in Linux command system. Downloading the materials for the tutorial is also instructed in this chapter. Though these tasks themselves are technical issues, we can experience through the tasks concrete examples that might be a trap for command-line-beginners to stall their learnings. Tips to avoid such stalling are given in the final section.

2.1 Minimum Guide for Linux Beginners

All the Linux operations [1] required in this book are summarized in the appendix chapter (Sect. 9, jumped from the end of this chapter). This section provides the minimal explanation for the set of commands, which are required to prepare Linux tutorial (beginners have to use Linux commands BEFORE they get the environment to learn the commands).

How to execute commands

When you are asked to execute

The original version of this chapter was revised: Belated corrections have been updated. The correction to this chapter is available at https://doi.org/10.1007/978-981-99-0919-3_14.

© The Author(s), under exclusive license to Springer Nature Singapore Pte Ltd. 2023,   31
corrected publication 2023
R. Maezono, *Ab initio Calculation Tutorial*,
https://doi.org/10.1007/978-981-99-0919-3_2

```
% cd
```

it means to type just "cd" on the terminal, and type enter key subsequently. Beginners sometimes tend to type starting with "%", but do not type "%". This "%" is called as "terminal prompt", which indicates that "This is the terminal input deck. Users are asked to input required commands after this symbol". "Type enter key!" is sometimes shortened as "enter!".

Grammatical structure of command

As an example like

```
% cd Desktop
```

the line takes the general structure as "% [command] [argument]",[1] separated by a space. Beginners sometimes forget to put the space between command and argument. Always be aware that "this is a command, this is an argument, and there should be a space between them" when you type.

The line would take further arguments like

```
% cp file01 file02
```

in a form as "% [command] [1st argument] [2nd argument]".

When it takes like

```
% tar -xvf linuxBasic.tar.gz
```

the block, "-xvf", works as the option specification for the command, taking the general form, "% [command] [-option] [argument]". All the components should be separated by a space.

Instruction for command executions

When the text asked to execute

```
% cd
% cd Desktop
```

it means that "Type 'cd', then type enter key. Wait until the prompt appears. Then, type 'cd Desktop'".

---

[1] See Acronyms for the usage of square brackets.

Tab completion

When you type, for example, '% cd Desktop', you don't have to type the whole long string, "Desktop", but instead, just type the first several characters like "Des" and then press Tab key by your left little finger. The system will then automatically fill in the rest of the string from the possible candidates. This is called **"tab completion"**. It is very important to get into the habit of using tab completion so as not to stall your motivation to learn this course (struggling with typing is the biggest obstacle for beginners). Always use tab completion and get into the habit of it as quickly as possible (Sect. 9.1.2).

How to indicate the location

The indication like "xxx/yyy" means that "'yyy'(file or folder) located under 'xxx'(folder)". Instead of using the term "folder", we use the term "directory", as "'xxx' under 'yyy' directory".

"cd" command

In the GUI, you can "visually click with the mouse" to move the location from folder to folder, but in the command line, you need to execute the command

```
% cd [destination directory location]
```

to move "your work location". For the correct way to specify the "destination directory location" argument, please refer to Sect. 9.1.3.

If you execute "% cd" without the argument, it means "take me back to the 'base location'". Beginners are usually easy to get lost on where one is working now. Do not be afraid, you can warp back home just by typing "cd".

"ls" command

This command is used to display a "list" of files and folders "visible" under the current work location. Whenever you execute any command on the command line, get into the habit of typing "ls" afterward. The habit corresponds to the behavior, "always check where you are working and what kind of view (files and folders) you have from your location". Often, beginners do not do this and continue to work, getting lost and executing commands in the wrong place and getting the error "the target file does not exist here".

"pwd" command

This command is used to check where you are working at the moment. When you type "pwd", the name of the directory where you are working is returned in the format "xxx/yyy".

2.2   Preparing Terminal Environment

Now, let's start building a terminal environment for Linux command line running on the RaspberryPi. The present section describes the steps to assemble the purchased equipments, and how to set up the terminal environment. Assuming the full-beginner, the explanation includes how to utilize keyboard shortcuts, what is the network installation, what is the super user, and how is the relationship between command line and GUI operations. Although the contents may be redundant for proficient users, it would be useful even for them, as an instructor, to understand what point should clearly be explained in advance for full-beginners to prevent them from stall. The usage explained in this section would be cited later frequently in the book.

2.2.1   Assembling RaspberryPi

The RaspberryPi purchased as a stand-alone device rather than as a kit suffices: you can use a kit, but it will be more expensive due to the extra accessories not required for the tutorial in this book. In addition to the main unit, you need a power adapter, a video adapter, an SD card (microSD), and an SD card reader. Approximate prices are given in the Foreword.

The flow of the procedure is as follows: The first thing is to prepare the SD card on which the Linux OS will be written using your PC (Win/Mac/Linux) with Internet access. Then, connecting the power supply, display, keyboard, and mouse to the RaspberryPi with the SD card inserted, turn it on to start up the RaspberryPi machine running the Linux OS.

Preparing SD card

Using your PC, connect your Internet browser to the website,

```
https://www.raspberrypi.org/software/
```

to get "RaspberryPi Imager" to be installed to your PC.

Next, insert the SD card into the SD card reader, and connect it to the PC. Make sure that the card is recognized correctly as a disk space. After confirming that, run the "RaspberryPi Imager" application on your PC.

When you start the application, you will be prompted to "Select OS" and "Select Storage for export". For OS selection, select

```
'Other general purpose OS' > 'Ubuntu'
    > '64-bit desktop OS for Pi 4 models with 4Gb+'
```

For "Storage for export", select the inserted SD card as recognized from the PC. Then, press "Write" button to execute writing.[2] When the writing is completed successfully, you can see a content on the SD card, named "system-boot".

Initial setting of RaspberryPi

Insert the SD card into the card slot of the RaspberryPi, connect the power supply, keyboard, mouse, and display, and turn on the power. After waiting around 2 minutes until the display responds, a panel appears shown as "System configuration". Select "English and Continue", and then "English (US) and Continue" to proceed.

Next, the setting on Wi-Fi connection will appear. Select "Connect to this network" by the radio button selection, specifying your own Wi-Fi with the password for it. Then, click the connect button to proceed.

Next, the panel appears asking "Where are you?". Select your location such as Tokyo and proceed. Then the panel showing "System Configuration" will appear. Configure it as follows (note the meaning of the square brackets is given in Acronyms)[3] :

```
- Your name              --> student
- Your computer's name   --> rp01
- Pick a username        --> student
- Choose a password      --> [yourChoice]
- Confirm your password  --> [yourChoice]
```

Then, select the radio button for "Require my password to log in" and click "Continue". You will see a screen that says "Welcome to Ubuntu" and wait for about 10 min to complete configuration.

The first login

Once the initial settings have successfully been completed, the login screen appears where you can login using the password you have just set. The first time you log in, you will be asked a number of questions, which you should answer as follows:

---

[2] It would take around 30 min.

[3] For "name = student" and "computerName = rp01" in the example, you can replace them with whatever you like, but for beginners who are not confident, it is strongly recommended to use "student/rp01" as specified for the time being. If you have changed "student/rp01" to your preferences, you need to replace "student/rp01" to those whenever it appears in the text. It has been our experience that most of the inquiries telling "I got an error and I don't know what to do" are due to forgetting the replacement.

```
- Connect Your Online Accounts --> 'Skip'
- Help improve Ubuntu          --> Either Yes/No
- Privacy                      --> Next
- You're ready to go           --> Done
```

Done the above setting, let's log out first: In the upper right corner on the screen, you will find a "power button" symbol. Click on it, then you will find an item showing "PowerOff/LogOut". Select "LogOut" to do it.

2.2.2 Getting Used to Shortcuts

Logged in the system again, find the browser (Firefox) icon in the upper left corner on the screen. Open the Firefox, and type something in the search deck [e.g., JAIST] to see if the browser is working with Internet connection. You can find the string, [Firefox Web Browser], in the upper left corner. Click it and then click [Quit] to close the application.[4]

Let's open Firefox app again. Then, with keeping to hold down the Ctrl-key, press the Q-key[5] (Such operation will be referred to as "Ctrl+Q" from now on). You will see that the application has been closed. To quit the app, beginners might use one's mouse to "click on the upper left display, select Quit, and click", but if you remember that "quit the app" can be done by "Ctrl+Q", you can perform the same action in one shot. These labor-saving operations are called **keyboard shortcuts**.

For beginners, keyboard shortcuts are often dismissed with the word like "I have to learn every single one?". But it's not something you learn in your head, but "learn by your fingers". Once you get into the habit of using keyboard shortcuts, your typing performance will increase at an accelerating rate.[6] The trick to learning keyboard shortcuts efficiently is to be aware of each **mnemonic**: e.g., Ctrl+Q, where Q means "quit".

Let's start Firefox again and browse any web page (e.g., JAIST). Then try "Ctrl+N" ("N" = new). You will get a new window opened. Then hit Ctrl+Q to exit the application. At this point, the dialog "Close tabs and quit?" will appear and two buttons "Cancel/Close tabs" will be displayed. Beginners might use the mouse to click "Close tabs", but do not take this style. Confirmed that the "Close tabs" button

---

[4] The first thing you need to learn on the computer system is always "how to quit".

[5] Ctrl-key is kept by your left thumb and hit "Q" with your right index finger. Beginners often try to use left little finger for Ctrl and index for "Q", enforcing everything within left hand, that's not good. When such an unnatural fingering becomes a habit, it will slow down the operation and prevent you from learning the shortcut.

[6] Even if you are at the beginner's level, many of you use "Ctrl+S" to save a document when you edit it using MS Word. Those who don't know "Ctrl+S" would use the mouse every time to save the document by moving it for "File > Save" on the screen. You can easily imagine how this will significantly slow down the operation speed. The "Ctrl+S"-users may feel like shouting "Don't use the mouse every time!".

is blue colored, hit the enter key. Then you can find that the application has quitted. The button shown in blue is the "default selection", meaning that if you just hit the enter key, it is equivalent to "clicking the blue button". This convention is commonly used even in Win and Mac, and for any applications. Knowing this convention can significantly improve the efficiency of not only this course, but also your daily work with the PC, by saving the time to find the mouse.

Open Firefox again, and then try Ctrl+T ("T" = tab). A new tab will be launched. Let's browse any different web pages shown on each tab [e.g., JAIST on one tab and "maezono group" on the other]. Then hit "Shift+Ctrl+PageUp (or PageDown)". You will see that you can go back and forth between the tabs without using the mouse. Then, try "Ctrl+T" again to get the third tab. Played with "Shift+Ctrl+PageUp (or PageDown)" to switch among tabs, then type "Ctrl+W" ("W" = window), and you will see that the tabs will disappear. Let's use this to reduce the number of tabs to just one.

Next, try "Ctrl+'+'" and "Ctrl+'-'".[7] You can adjust the size of the text displayed. As we explained so far, there are a lot of shortcuts. You can learn them by searching "Ubuntu Firefox shortcut" on Google. You don't have to memorize everything with your mind, but if you get into the habit of "searching for useful shortcuts" for frequently used operations and then "learning them by your finger (on typing)", you will gradually get use of it.[8] Note again that "Ctrl+Q", "Ctrl+Q", "Ctrl+S", etc. are not "just for this app, just for this OS", but common for almost all apps with the same mnemonic.[9] It is, therefore, well worth mastering!

2.2.3   Preparing Terminal Environment

After closing all of Firefox using the shortcut, click on the "Activities" button in the upper left corner of the screen. You will find the window shown as "Type to search", then type "terminal" in the window. Click on the "Terminal" icon that appears as a suggestion, and the Terminal screen will appear. Use "Ctrl+'±'" to adjust the font size as needed. Try "Ctrl+T" or "Ctrl+Q". You probably won't get any response, but instead, "Shift+Ctrl+T" or "Shift+Ctrl+Q" will work as you expect. This is often the case, some apps need additional "Shift+". For going back and forth between tabs "Shift+Ctrl+PageUp (or PageDown)" is working as common to Firefox app. Then you can exit the app by "Shift+Ctrl+Q".

Again, display the "terminal" icon from "Activities" (button in the upper left corner of the screen), and this time, try "right-click" on the icon. Select "Add to

---

[7] "Ctrl+'+'" means "Ctrl + Plus", and "Ctrl+'-'" means "Ctrl + Minus", respectively. Note that MinusKey directly shows "-" while PlusKey shows "+" with ShiftKey. The instruction "Ctrl+'+'" is therefore corresponding to the actual operation "Shift + Ctrl + PlusKey".

[8] Useful shortcuts can be taught to each other during a discussion, stolen from your colleague's operations, or found by accidental keyboard manipulation.

[9] In Linux, the "Ctrl" is simply replaced with "Cmd" in Mac and Win.

Favorites" and click on it. Then you will see that the Terminal icon has been added to the left side of the screen. From now on, you can launch Terminal by clicking on it.

Open the terminal and type "cd" and press enter.[10] This operation will be referred to as "**execute 'cd'**" and will be written as

```
% cd
```

Then, execute

```
% exit
```

You can see the terminal app has quitted. Here you can understand that the following operations are equivalent:

- Using the mouse to press "Quit App" (GUI operation).
- Using a shortcut, type "Ctrl+Q" (GUI operation).
- Quitting the application by executing the command "exit" (command line operation).

Then execute

```
% reboot
```

The RaspberryPi itself will be rebooted. As you can see, rebooting a PC, which used to be done with your mouse clicks, is now actually realized by a command operation. Once the RaspberryPi is up and starts running, try executing

```
% shutdown -h now
```

You will see that the RaspberryPi is now shut down.

After doing so, reapply power and start up the RaspberryPi again to continue the exercise.

2.2.4 Hierarchical Structure of Directories

For a beginner with only MS Windows experience, the most confusing part for the command operation is "I cannot figure out where I am" or "I can't find a file". The "click to open folder" operations commonly appearing in GUI interface do not exist anymore. We do not use this intuitive interface, but rather the old-fashioned "command and response" dialogue with the computer used in the command operations.

---

[10] The meaning of this command is explained in appendix (Sect. 9).

First, let's understand the "hierarchical structure of directories" by comparing it to the "GUI folder structure" that we are used to seeing, and how the same thing looks in command operations [the following assumes that you have read the "Minimum Guide for Beginners" (Sect. 2.1)].

On the left side of the screen, you will find the "Files" icon as the third from the top. Click on it, and a window will open. You will see a folder labeled "Desktop/Documents/Downloads/...".

Now, start up a terminal and execute "cd" and then "ls" on the terminal. What is returned on the terminal is a list of files and folders (→ Sect. 2.1), and you will see "Desktop/Documents/Downloads/...", which is the same as what we just saw in "Files".

Go back to the window opened by "Files". Right-click your mouse to see "New Folder". By using it, create a new folder named "test01". After doing so, go back to the terminal and execute "ls" again. You will see that a new folder named "test01" has been created there.

Next, go back to "Files", click on test01, go inside the test01 folder, and there create a new folder called test02. Then go back to the Terminal and execute "ls" to make sure test01 is there, and then execute "cd test01" [which means go inside test01 (→ Sect. 2.1)]. Executing "ls" again shows the contents inside the test01 folder, and then you will indeed see "test02" there.

This is the correspondence between "how the files are found on the GUI" and "how the files are found the command line". PC users used to operate by using the command line to handle the files as described above (up to around middle 80s), but this was too high a threshold to attract a large number of users. Then a more intuitive way of "clicking on the picture of a folder so that the contents can be seen as icons" has well been developed as GUI like Windows OS to get exploding number of PC users.

Then quit "Files" (as the GUI application) by "Ctrl+Q". For the terminal, type "cd" and then "pwd", and make sure that "/home/student" is returned, and continue with the rest of the exercise.

2.3   Installation of Required Tools

The following utilities are required in the present tutorials for file format conversion, visualization, etc. These will be installed from the Internet using the command "apt" known as **Package Management System**.

1. **Tools used to install other resources**: git.
2. **Visualization**: gnuplot.
3. **Text editor**: emacs.
4. **Simulation software**: quantumEspresso.
5. **Remote connection**: openssh-server.

When installed, "git" works as a command to obtain source codes distributed on the network. It should therefore be installed first, by using which you can download the example files for this tutorial.

"Quantum Espresso" is distributed in binary form ($\rightarrow$ Fig. 1.6) for use on the RaspberryPi. Therefore, no compilation is required in this course, but it is also possible to be compiled from the original source on the RaspberryPi.

The following sections describe the concrete steps for the installation.

2.3.1   Network Installation Using Sudo Command

Required tools are installed by downloading them from the Internet. In our environment, it is performed as

```
% apt install [application name]
```

> **! Attention**
>
> A network installation command like "apt" is officially called "**package management system**", which is "apt" for Ubuntu, "brew" and "port" for Mac, and "yum" for CentOS and Fedora. There are many types of package management systems, for which you cite Wikipedia to find further information.

Then, let's install "emacs", the editor we will use in this course:

```
% apt install emacs
```

There will appear error messages whinging "Permission denied". This rejection is because the installation of the application is "a fundamental modification of the system", and hence not allowed to be performed without a certain level of authorization. The authorized person allowed to perform such operations is called a **super user**. A user usually operates as a **general user**, not a super user. If you execute "**whoami**", you can find out as what username you are operating now. Executing it, the system will respond "student". Since the "student" is a regular user, we cannot install anything as it is. If you want to execute something "as a super user" from a regular user deck, you can use the command form,

```
% sudo [command]
```

Then, for the present installation,

```
% sudo apt install emacs
```

works. When the password will be asked, input the one you are using to login to RaspberryPi.[11] During the installation, you will be asked to enter "Y/N", for which enter "y" to proceed.

! Attention

You don't need to type "apt install" every time, it's the same string you typed before. Hit the up-arrow key, then the command you just executed will be displayed again, on which you can use the arrow keys to add or modify it as needed. It is very important to learn as early as possible how to avoid "lengthy typing" that can easily lead to typing mistakes. Typing mistake is the biggest cause for the beginners to lose one's motivation to master the computer system.

! Attention

It is quite frequent to find the beginners whinging that "I entered my password, but it won't to be input". It's not true, the input password is surely going through, but just "not displayed" for security purposes. Don't hesitate to continue typing your password and press enter.

2.3.2 Installing Operation

Then, let's install the utilities described at the beginning of this chapter in sequence as follows:

```
% sudo apt install -y git
% sudo apt install -y gnuplot
% sudo apt install -y openssh-server
```

In the above, the commands are executed with "-y" option (Sect. 2.1). Without this option, you will be asked "Y/N" during the installation process, as in the previous example. By this option, you can omit to respond each time.

! Attention

The above set of commands can be executed at once by

```
% sudo apt install -y git openssh-server gnuplot emacs
```

---

[11] On GUI, you may often be prompted to enter a password when installing software. The GUI is actually executing "sudo install ..." in the background.

However, it is safer for beginners to make it step by step to prevent from typing mistakes.

Then, install "Quantum Espresso" as

```
% sudo apt install -y quantum-espresso
```

Then type "cd" (going to home directory), and execute "pw.x" (the command to run "Quantum Espresso"). If the installation has been successfully completed, you will get

```
Program PWSCF v.6.7MaX starts on  6Aug2021 at 17:44:32

This program is part of the open-source Quantum ESPRESSO suite
...
Waiting for input...
```

Then type "Ctrl+C" to force quitting ($\rightarrow$ Sect. 9.1.3), followed by

```
% rm input_tmp.in
```

That's all for the required installations.

2.3.3   Remote Connection from Your Familiar PC

The contents of this chapter are not necessarily required if you decide working directly on the RaspberryPi. However, it is mostly assumed that the newly installed display and keyboard on the RaspberryPi will be placed in a cramped position and that the working environment will be less comfortable than the PC (Windows/Mac) you are used to. It is also important to establish a remote connection here in order to develop the habit of "taking notes on a familiar editor" as described in Sect. 2.5.1.

If your familiar PC and RaspberryPi are connected to the same Wi-Fi, you can follow the steps described in this section to remotely connect and control the RaspberryPi from your familiar PC with more comfortable typing environments for the continuing practices.

First, on the RaspberryPi, execute

```
% ip a
```

Then you will get the contents like,

```
% ip a
    1: lo: <LOOPBACK,UP,LOWER_UP> mtu ...

      ...

    3: wlan0: <BROADCAST,MULTICAST,UP,LOWER_UP> mtu ...

      ...

        inet 10.0.1.39/24 brd 10.0.1.255 ...

      ...
```

In the example above, "inet 10.0.1.39" can be found, and "10.0.1.39" is the IP address of the RaspberryPi (let's say "xxx.yyy.zzz.aaa" in the following).

If "openssh-server" has been installed as described in the previous section, you can remotely connect to the RaspberryPi by executing

```
% ssh student@xxx.yyy.zzz.aaa
```

from the terminal environment on your PC that you are familiar with.[12]

If your PC is a Mac/Win, please search for, e.g., "Windows ssh", to prepare the environment for using ssh connection from each PC. If the remote connection is successful, try repeatedly "exit" and "log in" again to become familiar with the ssh remote connection.

2.4   Setting Up Working Directory

2.4.1   Preparing Working Directory

Let's prepare a "work place" for the tutorials on your RaspberryPi. The preparation to "learn command operations" should be done using "command operations" before you learn, that's annoying. The operation here is then so-called "jargon operation"[13] for beginners. Type carefully to avoid mistakes (If you look at this section again after you have mastered Linux, you should have grown to understand what you were doing).

Execute

```
% cd
% mkdir work
% cd work
% pwd
```

---

[12] The first time you are asked Yes/No, then answer "Yes".

[13] Those like chanting a sutra or a verse that you don't understand, you just type it as it's written.

For the last "pwd", you will see

```
/home/student/work
```

This directory is abbreviated as

```
'/home/student/work' = '~/work'
```

Namely, the tilde ($\sim$) abbreviates the home directory '/home/student'.

2.4.2   Obtaining Tutorial File Sets

A set of files [introductory exercises for Linux operation (Sect. 9), a set of input files for tutorials starting from the next chapters] are available on web. This can be downloaded by using the "git command" as follows.

Execute the following to move to the working directory:

```
% cd
% cd work
```

Then, execute

```
% git clone https://github.com/ichibha/setupMaezono.git
```

You will get

```
    Cloning into 'setupMaezono'...
    remote: Enumerating objects: 19, done.
    remote: Counting objects: 100% (19/19), done.
    remote: Compressing objects: 100% (15/15), done.
    remote: Total 19 (delta 1), reused 18 (delta 0), pack-reused 0
Unpacking objects: 100% (19/19), done.
```

to complete the download. Executing "ls", you will find a directory named "setup-Maezono", in which the set of files are located.

Next, move to "~/work/setupMaezono". Execute "ls" and you will find a file called "linuxBasic.tar.gz" inside the folder. This is a compressed file including the set of tutorial files in a single file. The extension, "tar.gz", indicates that it is a compressed file.[14]

---

[14] There are a variety of compression formats, some commonly used on Win, some on Mac. The "tar.gz" is a widely used format on UNIX/LINUX. It is sometimes referred to as "tarball it and email it to me".

Being sure to use tab completion, and being careful with spaces, execute

```
% tar -xvf linuxBasic.tar.gz
```

"tar" is a command for compressing/decompressing files, where the argument "setup-Maezono.tar.gz" is the name of the file to be compressed/decompressed. The option "-xvf" in between expresses "use in decompression mode". Again, remember the command format "**(command) -(option) (argument)**" (each separated by a space).

Completed the decompression, execute

```
% pwd
   /home/student/work/setupMaezono
% ls
```

If you can see "us" or "bash_alias", the decompression has turned out to be completed successfully.

! Attention

The above operation

```
% pwd
   /home/student/work/setupMaezono
```

is made in order to announce to the reader that "we are about to work in this directory location". This kind of consideration is very important when asking questions or giving instructions for operations remotely. Most of the errors reported by beginners are that they are not working in the right place.

The file set includes the input files in the tutorial. For beginners, one may inevitably make typing mistakes to get errors, or one may get errors due to technical problems due to character codes (such as newline character). To avoid such errors, the readers can use the downloaded input files for comparing with what you typed or for cut-and-paste purpose.

2.4.3 Setup Alias File

The commands sometimes take longer and complex by the additional options and arguments with longer strings. If such a longer typing appears frequently, it's better to "register" it on your "dictionary for the commands". For example, if the command "% ls -CF" is frequently used, you can register it as "l='ls -CF'" on some "dictionary file" called **alias** file. Once the alias file is activated, then you can execute "l" playing equivalent to "ls -CF".

For this tutorial as well, an alias file is prepared and provided in the downloaded "linuxBasic.tar.gz". The file contains aliases for frequently used commands as well as the settings for **command path**.[15] To make the tutorials go smoothly, execute the following operations to activate the alias file.

Execute

```
% cd
```

to go to the home directory. Type the following string using tab completion, but without typing enter key:

```
% ~/work/setupMaezono/bash_alias
```

(Do not press Enter. If you do, press the up-arrow key to recover.) Then, "Ctrl+A" to move the cursor back to the beginning, and then add "echo 'source'" as

```
% echo 'source ~/work/setupMaezono/bash_alias
```

Then, "Ctrl+E" to move the cursor to the end of the line and add """ as

```
% echo 'source ~/work/setupMaezono/bash_alias'
```

Then, press enter key. The above "echo command" will return a string enclosed in single quotes to the terminal. Make sure the returned string shown as "source ~/work/setupMaezono/bash_alias" with no typos [if there is a typo, redo "% echo ..."], then type "pwd" to confirm that you are at your home directory.

Then, use the up-arrow key to display the above echo command again, and add ">> .bashrc" to the end of it as

```
% pwd
   /home/student
% echo 'source ~/work/setupMaezono/bash_alias' >> .bashrc
```

Be careful about the "." (dot) in front of bashrc. For a detailed description of the meaning of "file names starting with a dot" (**invisible file**), as well as the detailed meaning of what we are doing here, please refer to the appendix chapter (Sect. 9).

To confirm that the above settings have been correctly applied, see the contents of .bashrc file as

---

[15] By knowing the location of the directory in advance, in which frequently used commands are located, the system can execute the commands with less typing, otherwise fully specified location (longer string) is required to be typed. The location is called "command path", which is written in the alias file and activated.

```
% cat .bashrc

   ...

   source ~/work/setupMaezono/bash_alias
```

As above, if you find "source ..." at the end, the setting has successfully completed. Completed it, then whenever you login the system, the alias file (bash_alias) we distributed is automatically activated.

To test it, exit from the terminal and start the terminal again to type

```
% cd
% c
   student/
```

If you see "student" as above, then the automatic activation of the alias file is successfully setup.

2.5 Tips for Successful Learning

2.5.1 Cease Habit of Handwritten Notebook

Before going to the tutorials, the author note that please avoid writing down what you are taught in your notebook by pencils. Many beginners including me tend to take note by handwriting, and then to refer notebook by hand each time when they try to remember what to type. This is far ineffective making them reluctant to work on the command line. This is the main reason why most of beginners, especially those joining from experimental group, fail to master the skill. In particular, the students who have established successful experiences (including me) in course works (classroom with blackboard!) tend to be difficult to move away from that style. However, handwritten notebooks are the most inappropriate way for simulation tutorials. The arguments for the command, the directory where the command executed, etc. vary from time to time, depending on the situation. Beginners tend to try to write down everything on one's notebook, but the detailed information like directories or argument used at that time has no meaning in another situation. Such beginners who rely much on notebooks encounter more errors because they execute commands without considering the situation in exactly same way as what is written in one's notebook (with improper arguments for the situation).

You can view the history of the commands you have executed by using the up-arrow key or the "history" command. It is important to get into the habit of "taking notes on the editor" by using these commands and cut-and-paste them on the text editor such as "Notepad" application. It is quite likely for you being not familiar with the Notepad app on your RaspberryPi (the author too!). If so, you can connect to the RaspberryPi remotely from your familiar PC (Win/Mac) and take notes on your PC, as explained in Sect. 2.3.3.

2.5.2 Tips to Prevent Discouraging Your Learning

This section is intended as a reminder for Linux beginners, so proficient users may skip it. Beginners are recommended to move on to the appendix chapter (Sect. 9) after reading this section. "Minimum Linux skill required for this tutorial" is provided in the appendix using the set of materials downloaded in Sect. 2.4.2.

To avoid stalling of your motivation to learn, please make sure the following items again. As explained in Sect. 8, the most common cause of stalling is a lack of proficiency in the following **technical** matters that can simply be acquired with one's effort (effort to get habit).

1. **Spatial awareness on directory structure**
   Always type "ls" or "pwd" to keep track of where you are, even if you are not told to do so. Most of the case when a beginner says "I got an error" is because of a mistake like "You're executing the command at wrong location". When you execute commands, be always aware of where you are working. This is the first step in getting out of the earliest beginners who type commands like chanting a sutra without thinking logically.

2. **Touch-typing**
   In order to grade up your PC from "a tool for MS Word" to "a desk for thinking", you should definitely learn touch-typing. It may take a couple of months, but you will surely be able to make up for it in comfort (The author couldn't do touch-typing until about 26 years old!). In simulation research, it requires a lot of typing. If you don't remove the **psychological barrier** of facing the keyboard, your motivation will easily stall.

3. **Tab completion**
   As repeatedly explained, no need to type the entire string, but just type up to the middle of the string and press the tab key. Then, the system will automatically fill in the rest of string for you. Without tab completion, the frequency of errors due to mistyping will increase dramatically, and you will gradually grow tired of it and lose your motivation to learn. Tab completion is also very effective for "daily use" of PC, such as email, document editing, and Excel handling. It surely improves your life efficiency several times over. Strengthen your "little finger of the left hand" to seek tab key. (It is recommended that you tap the desk or your knee with the little finger whenever it's free. Useful also for playing the guitar.)

4. **Try to memorize it**
   Try to memorize the commands as possible as you can. You can do this if you try. A little hard work in the beginning can save you years of work later.

5. **Keyboard shortcuts**
   There are few users who click the mouse to save a file every time they update a document. It is easy to imagine how much convenience is lost when using "Ctrl+S" is prohibited. With a little harder effort, you can learn the other shortcuts. It surely saves you years of work later.

6. **Ensure use of "alias"**

   Instead of "not memorizing", but "minimizing what you need to memorize" is the important concept. No need to memorize all the complex combinations of commands and arguments. It is possible to "register" the combinations as aliases and execute them as a new single command. Alternatively, you can create a command using the alias that displays the usage of the command for reminding. Once you make it a habit to register everything in alias file for frequently occurring tasks, you will be able to start up every task much faster and more efficiently. This is also important to prevent yourself from losing motivation.

7. **Strong spirit not to leave things to others**

   At more senior positions with supporting staff, it gets mode difficult to master new things (even as a reminder to the author). To learn something new, one needs to try it out by one's own hands. While professors are saying that "simulations are too difficult to understand", their graduate students are learning and mastering it in a few months. It's not that it's difficult, it's just that they have lost the strong spirit to use their own hands.

Reference

1. "Linux Command Line and Shell Scripting Bible", Richard Blum, Christine Bresnahan, Wiley (2021/1/13) ISBN-13:978-1119700913

Sequence of Computational Procedure

3

Abstract

In this chapter, we will follow the whole flow of the kernel calculation, which is to perform electronic structure calculations for a given material structure and to check the fundamental quantities obtained. The flow of the calculation is a series of "preparing basic input files", "self-consistent calculation", and "calculation of electronic structure". To understand the structure of the sequence first, we try running a simulation program of the density functional theory using the "already prepared and given input files" and try processing the output to obtain the band dispersion diagram, etc. The following points are the check points for understanding this chapter:

- **Difference between [making "dough"] and [using "dough"]**
  The most beginners tend to focused on using the "dough" before they could create right "dough".
- **Preparing input files**
  What kind of input files should be prepared and how they are prepared?
- **Executing calculations**
  How to run calculations using command line, and how to save the results.
- **Confirming results**
  How to confirm whether a calculation has been completed with proper behavior? In particular, we learn how to use a plotter easily for this purpose.
- **Evaluating electronic structure**
  What is the significance of calculating the density of states and dispersion plots? How to calculate and display them?
- **Simulations as a quick check**
  How to confirm if the "dough" is created successfully?

The original version of this chapter was revised: Belated corrections have been updated. The correction to this chapter is available at https://doi.org/10.1007/978-981-99-0919-3_14.

R. Maezono, *Ab initio Calculation Tutorial*,
https://doi.org/10.1007/978-981-99-0919-3_3

Detailed explanations of each process are given in the following chapters. In this chapter, please proceed with putting your attention to understand the sequence of tasks first.

3.1   What Is the Self-Consistent Calculation

In the "workflow" mentioned at the beginning of this chapter, "preparing the input file" is fine to understand, but there are two calculation processes, namely, "self-consistent **calculation**" and "electronic structure **calculation**", and the relationship between these two calculations is difficult to understand for a beginner. One might be curious about the sense of the term "self-consist", but for the time being, just think of it as a name of a task in the procedure, it will be explained in detail in the later chapter (Sect. 5.3.5). The self-consistent calculations aim at constructing "self-underlineconsistent field", so it is sometimes called **SCF calculation**.

Now, about the relation between the two calculations, "self-consistent **calculation**" and "electronic structure **calculation**", it can be metaphorically described as the relationship between

1. Making the bread dough itself (fully matured and fermented).
2. Baking bread from the finished dough.

There are two ways to say "bake bread", but

- **2a**: To check if the bread dough is done properly, bake a standard "tin bread" first to make sure it is done properly and
- **2b**: The completed dough is used to make a variety of bread products

are for the different purpose. It is "2a" that we will deal with in this book mainly. This is one of the point where **beginners are likely to be confused** (Fig. 3.1).

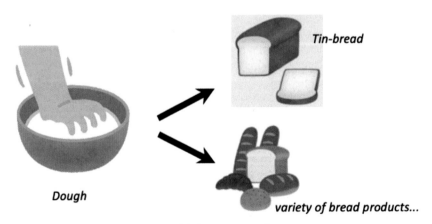

Fig. 3.1 Bread dough used for tin-bread and variety of breads

If I start to teach "2a" without explaining with the phrase "to check if the dough is done properly", students sometime whinge that "Why we have to learn how to bake tin-bread, I join the course to learn how to make fancy bread for my future with my shop". If you skip the step of making sure the dough is well finished, you will end up with a dough that is not mature enough to make fancy breads. Along this metaphor, "making a well-fermented bread dough" corresponds to "obtaining a well-converged electronic state (wave function) by calculation".

Concrete image of the "convergence" will be explained in detail in the later chapter (Sect. 5.3.5). When consulted by beginners whinging "My result looks strange", most of the cases are due to the fact that they have proceeded with the calculation to evaluate various properties using an "electronic state" that is not sufficiently converged (and they do not know how to check it). Too focused on desired properties (fancy bread), the importance of "properly preparing the dough" (SCF calculation) is not well recognized. The most difficult part is actually to prepare "properly matured dough", and once this part is cleared, "baking various breads" itself is not so difficult to understand.

3.2 Preparing Input Files

3.2.1 How to Prepare Structure/Geometry Files

In this course, we will deal with a series of calculations on the subject of energy comparisons between the four structural isomers of silicon oxides: α-quartz, β-quartz, α-cristobalite, and β-cristobalite ($\rightarrow$ Fig. 3.2).

As mentioned in Sect. 1, the most essential input to be given to the ab initio calculations is the **geometry** which specifies which elements are located in which spatial positions. The first task for a user is therefore to provide the geometry as an input file. In the present case, the crystal structures for the four polymorphs shown in Fig. 3.2 should be specified in the file.

Nowadays, crystal structures over existing materials are well organized and provided in database as an accumulated knowledge of mankind. In the past, users were required to look up and understand the target structure described in books or original papers, and then need to edit a file to describe the structure as a numerical format following each convention that differ from application to application.

This step used to be quite a tedious task. Users had to first learn the vast amount of knowledge on the crystal structure specification as described in the textbook of Solid State Physics [1], and then read through the manual of each application to become familiar with the input keywords and its format of geometry specification. This part is a sort of "death valley" in any tutorial workshop, taking a lot of time and making participants get bored to reduce their motivation. Each application has its own input formatting, hindering such researches switching over different applications even for experts. Moreover, there were also a lot of different formats for the database, which made things even more tricky.

As time has gone on, however, the database format has become standardized to "**cif format**" for the most part. Each application only need to provide a utility that can convert the cif format to its input format (Fig. 3.3). The process that used to be the most tedious for beginners is now well proceduralized. Though, for more professional works, one needs to eventually acquire the skills to adjust the crystal structure in detail, you can start from similar structures obtained from a database existing on the network, and then tune them a bit.[1] Even for the tuning process, you can adjust the atomic positions using the mouse on the visualization software, instead of manually typing in text as in the past.

3.2.2 Obtaining Structure Files

The four crystal structures shown in Fig. 3.2 can be obtained from the "American Mineralogist Crystal Structure Database (AMCSD)" database.

> **! Attention**
>
> The following browser operation needs to be done directly on the RaspberryPi. If you are operating it remotely from your familiar PC (Sect. 2.3.3), please switch to

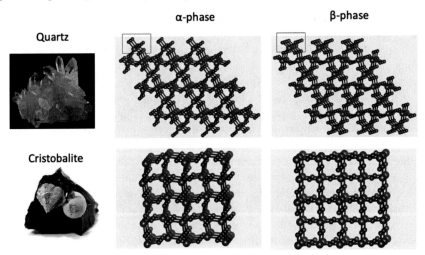

Fig. 3.2 The four crystal structures treated in this course are shown. The same SiO_2 composition can result in different materials depending on how the crystal lattice is assembled. The four consists of the combination, [quartz/cristobalite]⊗[α-/β-phase]. The blue ball (large) is silicon and the red ball (small) is oxygen. The structures of the two quartz phases, α and β, appear identical at first glance, but looking carefully at the area enclosed in red, you can see that red balls surrounding the blue balls are more asymmetrical in the α phase [pictures taken from Wikipedia]

---

[1] In the past, one had to edit the structure from scratch all by hand in a text editor, following the complicated format of each application.

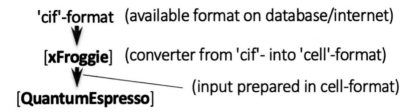

Fig. 3.3 "Quantum Espresso" program package used in this course takes the "cell" format as input to specify the geometry. For the geometries taken from database are written in "cif" format, and hence need to be converted to "cell" format using a utility called "xFroggie" (web application)

login directly to the RaspberryPi now (it is reminded later at the point where you can get back to the remote connection).

The Quartz structure can be obtained from

```
http://rruff.geo.arizona.edu/AMS/minerals/Quartz
```

(change Quartz to Cristobalite to get the Cristobalite structure).

> **! Attention**
>
> From here, you need to connect to the specified website several times and download the file. To prevent from mistyping for lengthy URL specifications, we have prepared a text file in ~/work/setupMaezono/memo.txt that contains the URL. Use the right mouse button to cut and paste the URL. Also, by hovering your mouse over the URL link, an underscore will appear, and when you right-click, there will be an "Open link" selection, which you can click to launch Firefox.

Looking at this page,

```
American Mineralogist Crystal Structure Database
45 matching records for this search.

Quartz (# This is the starting line of each information)
Download hom/quartz.pdf
Levien L, Prewitt C T, Weidner D J
Download am/vol65/AM65_920.pdf
American Mineralogist 65 (1980) 920-930
Structure and elastic properties of quartz at pressure
P = 1 atm
_database_code_amcsd 0000789 (#Here is the identification number!)
...

Quartz (# This is the starting line of each information)
...
```

```
P = 20.7 kbar
...
```

you will see that there are 45 blocks starting with "Quartz" in response to the "45 matching records" at the beginning, each with a link to download the structural data. Each item has an "identification number" such as "_database_code_amcsd 0000789". You may be wondering, "Why are there 45 structural data for the same structure?". This is because there are numerous variations in reports, such as differences in the year of the experimental identification or the temperature at which it was measured.

In this course, we use "α-quartz/0015462" and "β-quartz/0018071". Using "Ctrl+F" (search function of Firefox), you will get a entry window that appears in the lower left corner of the browser. Enter "0015462" in the window, and find the location of the corresponding item. Within an item, there are several links,

```
Download AMC data (View Text File)
Download CIF data (View Text File)
Download diffraction data (View Text File)
View JMOL 3-D Structure (permalink)
```

By clicking on "Download CIF data", the cif file will be downloaded. It can be found as a file named "AMS_DATA.cif" in the "Downloads" folder, which is accessible from "Files" application by launching its icon located at third from the top on the left side of the screen, as mentioned in Sect. 2.2.4). As explained in the subsection, the file is also accessible via terminal as[2]

```
% cd
% cd Downloads/
% pwd
     /home/student/Downloads
% ls
     AMS_DATA.cif
```

Let's save the downloaded file under the directory, "~/work/structure", generated as

```
% cd
% mkdir -p work/structure
```

Then, execute

```
% cd work/structure
% pwd
     /home/student/work/structure
% mv ~/Downloads/AMS_DATA.cif  .
```

---

[2] Be sure why "% pwd" is executed as explained at "Attention" in Sect. 2.4.2.

```
% mv AMS_DATA.cif quartz_alpha.cif
```

to save the file as renamed "quartz_alpha.cif".

Similarly, obtain the structure file with the id "0018071" from

```
http://rruff.geo.arizona.edu/AMS/minerals/Quartz
```

and save it as "quartz_beta.cif". Then prepare the files for Cristobalite, "α-Cristobalite/0001629" and "β-Cristobalite/0017665" obtained from

```
http://rruff.geo.arizona.edu/AMS/minerals/Cristobalite
```

to save them as "cristo_alpha.cif" and "cristo_beta.cif" to the same directory.

Once successfully stored them, check each file size to make sure that the download has correctly been done.

```
% pwd
    /home/student/work/structure
% ls -l
  total 20
  -rw-rw-r-- 1 student 1300  8 5 17:11 cristo_alpha.cif
  -rw-rw-r-- 1 student 4830  8 5 17:12 cristo_beta.cif
  -rw-rw-r-- 1 student 1058  8 5 16:49 quartz_alpha.cif
  -rw-rw-r-- 1 student 1056  8 5 17:05 quartz_beta.cif
```

Putting the option, "-l", to the ls command, the file capacity will be displayed (the number displayed such as "4830" is the capacity). Since the name of the downloaded file was always the same ("AMS_DATA.cif"), a mistake is likely to overwrite accidentally the previous "AMS_DATA.cif" that you forgot to delete just before and giving it a wrong title. It is highly unlikely that the file capacities will match, and so they should be different from each other. As such, the above check using "ls -l" can be a quick confirmation that there is no such mistake.

> **! Attention**
>
> From here, you can go back to connecting remotely from your familiar/comfortable PC.

3.2.3 Conversion of Structure Format

The next step is to convert the "cif" files obtained from the network into the "cell" format that can be read by "Quantum Espresso" using a web-based converter called "xFroggie" (Fig. 3.3).

Visiting the website,[3]

```
https://xfroggie.com
```

you can see the list shown as

```
-Convert crystal structure file format
-Generate k-path for band structure plot
-  ...
```

Clicking the upper most hypertext, 'Convert crystal structure file format', you can see the list,

```
1. Select file to read geometry
2. Select file formats of source and destination
```

By clicking the "Upload"-button below the text "1. Select file to read geometry", you can select a file to be uploaded and converted. Finding the location where "quartz_alpha.cif" is saved and choosing it to be uploaded.

Then, under the text "2. Select file formats of source and destination", we choose

```
'Source'='CIF-file'
'Destination'='Quantum espresso in file'
```

By clicking "Convert"-button, one will see the contents given in the "cell" format in the window below the "Download"-button. Confirmed the contents as properly given, click the "Download"-button, then the file "quartz_alpha.espresso-in" will be downloaded in the Download folder.[4]

When you see the contents shown in the window as given in "cell"-format, you can find the strings such as "ibrav = 0" and "CELL_PARAMETERS angstrom". Detailed explanations on these meanings can be found in the web manual of Quantum Espresso [2]. This part corresponds to a format specifying a crystal geometry. There are several different ways to specify the geometry. The simplest way is to give the position of the atoms as (x, y, z)-coordinates, but even for that, there is a difference in whether the unit is in angstrom or in atomic unit (Bohr). Another way is to specify first that the lattice is, e.g., BCC (body-centered cubic), and then to provide the lattice parameter and to give atomic positions as the ratio to the lattice parameter along the implied lattice vectors (called fractional specification). Even within this way, there are two different ways to specify the lattice vector, Conventional or Primitive, which causes confusion for beginners. In practice, it is sometimes found the trouble with strange

---

[3] The referenced URLs may have changed, so please refer to the latest link information given in the text file ∼/work/setupMaezono/memo.txt included in the downloaded set of materials.
[4] The location of the Download folder is the same as that explained in Sect. 3.2.2.

results due to the mistake about these Conventional/Primitive or Angstrom/Bohr confusion.

Then, let us locate the downloaded cell file "quartz alpha.espresso-in" into the working directory, ~/work/structure/,[5]

```
% cd
% cd work/structure
% pwd
      /home/student/work/structure
% mv ~/Downloads/quartz_alpha.espresso-in.
% mv quartz_alpha.espresso-in quartz_alpha.in
```

You can confirm the contents of it as

```
% more quartz_alpha.in
      &CONTROL
      /
      &SYSTEM
      ntyp  = 2
      nat   = 9
      ibrav = 0
      ...
```

This is the input in cell format.

Taking the similar procedure described above, please prepare the cell formats for quartz_beta and cristo_alpha converted as "quartz_beta.cif"→"quartz_beta.in" and "cristo_alpha.cif"→"cristo_alpha.in". For the rest structure, "cristo_beta.cif", we have to make a correction to a bug in the database in advance to the file conversion. Please open the "cristo_beta.cif" by emacs (use the command "emacs -nw [filename]"), and correct the following part:

```
    Before: \_symmetry\_space\_group\_name\_H-M 'F d 3 m'
Corrected: \_symmetry\_space\_group\_name\_H-M 'F d -3 m'
```

After this correction, please prepare "cristo_beta.in" taking the same procedure to generate other cell formats.

By the procedures above, we have prepared all structure files in the cell format. The files, however, only include geometry information, not the computational conditions for the simulation to be run at all. To add such information, let us edit the files to add the required keywords as given below[6]: Starting with "quartz_alpha.in",[7]

---

[5] Be sure to understand the purpose of "% pwd" as explained in Sect. 2.4.2.

[6] To avoid typos, the input files used in this course are included as text files in the downloaded materials as introduced in Sect. 2.4.2. You can copy them, or "cut and paste", or for comparing to find mistakes using "diff" or "sdiff" command. These files are stored with the same file name under "~/work/setupMaezono/inputFiles/".

[7] Be sure to remember why "% pwd" is executed, as explained in Sect. 2.4.2.

```
% pwd
   /home/student/work/structure
% emacs -nw quartz_alpha.in
```

In the editing deck, insert and modify the lines as instructed below with the comment tags "!":

```
&CONTROL
   calculation = 'scf'            ! added
   restart_mode= 'from_scratch'   ! added
   prefix='qa'                    ! added
   outdir='out_qa'                ! added
   pseudo_dir='./'                ! added
/
&SYSTEM
   ...
   occupations = 'fixed'          ! added
   ecutwfc = 50                   ! added
   ecutrho = 400                  ! added
/
&ELECTRONS
...
ATOMIC_SPECIES
   Si   28.085 Si.pbe-n-kjpaw_psl.1.0.0.UPF ! modified
   O    15.999 O.pbe-n-kjpaw_psl.1.0.0.UPF  ! modified

K_POINTS {automatic}              ! modified
2 2 2 1 1 1                       ! added

...
```

Perform the same modifications to the rest input files, "quartz_beta.in", "cristo_alpha(beta).in".

Then, in the next section, we will prepare the pseudopotential file, which is another input that should be put on the working directory.

3.2.4 Preparing Pseudopotentials

It was mentioned that the most fundamental input to be provided is the geometry of the elemental species (where they are placed). For example, taking a carbon dioxide molecule CO_2, the electrons in the system form their "electronic state" as the consequence to be influenced by the interactions from each nucleus as **atomic potential**. The electronic state is determined by the governing equation (Schrodinger equation), under the given atomic potentials located at the geometry. This is what the simulation package like Quantum Espresso is solving by the computation, described as "electronic structure calculation". Though the detailed explanations will be given later in Sect. 5.5, the rough understanding required at this stage is that the "effect of the nuclei array" as perceived by the electrons is given as a numerical potential, called the **pseudopotential** [3].[8] Therefore, pseudopotentials exist for each element, such as "pseudopotential for carbon" and "pseudopotential for oxygen", and it is necessary to prepare these for all elements in the target system.

In this tutorial, we work on silicon dioxide solids, so we need Si and O pseudopotentials. How to generate pseudopotentials is a highly specialized area developed by experts. As the outcome from such specialized field, numerical data for pseudopotentials are usually available on the web.

They used to be available from the Quantum Espresso web page, but are no longer available.[9] Therefore, the pseudopotential files used in this tutorial are distributed in the set of materials downloaded at Sect. 2.4.2.[10]

```
% pwd
    /home/student/work/setupMaezono/inputFiles/pp
% ls
    O.pbe-n-kjpaw_psl.1.0.0.UPF   ...
    O.pbe-n-rrkjus_psl.1.0.0.UPF  ...
    Si.pbe-n-kjpaw_psl.1.0.0.UPF  ...
    Si.pbe-n-rrkjus_psl.1.0.0.UPF ...
```

You see four files ending in "...UPF". The files starting with "O" are pseudopotentials for oxygen atom, while those starting with "Si" are pseudopotentials for silicon atom. **Variety of pseudopotentials are available for the same element** depending on how to generate them using different scheme [3]. We will explain the variations and how they should be selected later in Sect. 5.5. In this tutorial, two types of pseudopotentials are prepared for each of Si and O in order to verify "the difference in results depending on the choice of pseudopotential" later.

Let us look at the content of a pseudopotential file, for example, "O.pbe-n-kjpaw_psl.1.0.0.UPF",

```
% pwd
    /home/student/work/setupMaezono/inputFiles/pp
% more O.pbe-n-kjpaw_psl.1.0.0.UPF
    <UPF version=''2.0.1''>
    <PP_INFO>
    Generated using ''atomic'' code by A. Dal Corso  v.6.3
    ...
```

which seems to be an endless list of numerical data. This is the substance of pseudopotentials as numerical data.

---

[8] Why the word "pseudo" is used is understood through the explanation given in Sect. 5.5.

[9] It seems to be available again at the time of composing the present book.

[10] Directory name "pp" abbreviates pseudopotential.

Then, please copy the pseudopotential files to the working directory as[11]

```
% pwd
    /home/student/work
% mkdir pp
% cd pp
% pwd
    /home/student/work/pp
% cp ~/work/setupMaezono/inputFiles/pp/* .
% ls
    O.pbe-n-kjpaw_psl.1.0.0.UPF    Si.pbe-n-kjpaw_psl.1.0.0.UPF
    O.pbe-n-rrkjus_psl.1.0.0.UPF   Si.pbe-n-rrkjus_psl.1.0.0.UPF
```

Confirmed that the four pseudopotential files are shown by the "ls" command, then proceed the next section.

3.3   SCF Calculation

3.3.1   Preparing for Calculations

We have now prepared all input files necessary to run SCF calculation using Quantum Espresso. Create a directory named "~/work/01tryout" and copy two pseudopotentials under "~/work/pp", namely, "Si.pbe-n-kjpaw_psl.1.0.0.UPF" and "O.pbe-n-kjpaw_psl.1.0.0.UPF", to the "01tryout" directory. Then, copy "quartz-alpha.in" as "scf.in" into 01tryout.[12] We now have all the input files needed for the calculations under "01tryout".

Note that it is important practice to prepare a **working directory** and **copied** the necessary input files to it to perform the calculations, leaving the original files saved at the repository directories (such as "pp"). Another important practice is to execute the calculation as

```
% cp [input file with meaningful name] scf.in
% [command] < scf.in > scf.out
% mv scf.out [output file with meaningful name]
```

so that the input/output file has the unified name when the command executed. This manner well fits to further automations in which the "[command] < scf.in > scf.out" part is called repeatedly in the loop construction. If the input/output takes different names every time you executed, it is difficult to be implemented into the loop.

---

[11] Be sure to understand the purpose of "% pwd" as explained in Sect. 2.4.2.

[12] If you are well able to perform these instructions written in English using the commands like "cp", "mv", etc., you are no longer a Linux beginner.

! Attention

Since reproducibility is important in science, it is important to note that I/O (input/output files) should be kept as "valuable research logs". At the same time, however, it is necessary to save file space. As such, only the minimum set of I/O (required to reproduce the result) is kept while other unnecessary/secondary files to be deleted to avoid flooding the file space.[13]

Now the required input files are ready. The meanings of each parameters in this file will be explained in detail in the next chapter, but for the time being, let's just say that "the teacher has prepared the input file" that surely works. Then let's proceed to experience the overall flow of the operation.

3.3.2 Executing SCF Calculations

Now, it is time to run the SCF calculation. Even though your parameter settings are basically prepared by your "teacher", you can be impressed at this stage because a most beginner has downloaded, processed, and prepared input files by oneself. The command to perform the calculation was "pw.x" (Sect. 2.3.2), so it would be executed as "% pw.x < scf.in > scf.out". However, here, we do it with[14]

```
% pwd
    /home/student/work/01tryout
% pw.x < scf.in | tee scf.out
```

The reason for not using "% pw.x < scf.in > scf.out" is because, in this form, the output text messages are redirected[15] only to the file "scf.out", not displayed on the screen, making us impossible to know what's going on. The **tee** command appeared above solves this inconvenience. The form "[command] | tee [file name]" makes the output text be written into the specified file as well as displayed on the screen.

The calculation would take about 6 minutes.[16] Getting all standard output completed successfully, the prompt should come back right after the text,

---

[13] Without teaching this to a new student, I once experienced the accident that he could not reproduce any of his results because he deleted all I/O files, although the resulting good-looking graphs were ready after a while of research.

[14] For the redirect ">" and pipe "|", see Sect. 9.2.2.

[15] This is just what the word "redirect" means.

[16] A normal PC would take less than a minute, being much faster than RaspberryPi with relatively poor computing power. On a faster processor, it just takes a few seconds. However, "6 min" is amazing to be achieved by a processor with no fan!

```
PWSCF        :    5m11.29s CPU    5m34.65s WALL

This run was terminated on:  17:53:16    6Aug2021
=--------------------------------=
JOB DONE.
=--------------------------------=
```

Make sure that the contents are written to the "scf.out" file by using "more". This file is called **standard output file** (a copy of the standard output texts).

After the calculation is complete, the contents inside the directory should be as follows:

```
% pwd
    /home/student/work/01tryout
% ls
    ...[pseudo potentials]
    out_qa/  scf.in  scf.out
```

In addition to the standard input/output "scf.in/scf.out" (these should be kept as the "research log"), we can see that a new directory named "out_qa/" has been created. The directory name is in response to the specification of 'outdir='out_qa" specified in the input file "scf.in", as prepared in Sect. 3.3.1. This directory contains all information obtained as a result of the convergence that is used for every post-processing, called the **save directory**. The save directory is the "bread dough" required for further analysis. Although it is a valuable result obtained through the hard work of convergence (it was 6 minutes, but sometimes it might be a week even using powerful supercomputer!), it is large in size, so it is customary to delete it to save file capacity when the project has completed with all the post-analysis done. What should be kept as the "log" are only the standard input/output file, which is the "summary of what calculations were performed and what results were obtained" (leave the recipe but not the food itself). However, **do not delete "out_qa" at this time**, since the save directory will be used in the practice in the sections that follow.

3.3.3 Checking the Results

If you read through the standard output file, you will be able to understand whether the calculation finished successfully or not. It is, however, not very practical to read through such a large amount of content every time. Usually, one can check the result as

```
% pwd
    /home/student/work/01tryout
% grep 'total energy' scf.out
    total energy           =      -389.09239119 Ry
    total energy           =      -390.91105457 Ry
    ... (omitted)
    total energy           =      -391.13886726 Ry
!   total energy           =      -391.13886681 Ry
    The total energy is the sum of the following terms:
```

Here, an important command "**grep**" appears [4], used as

```
% grep '[string]' [target file]
```

This command extracts and displays only the lines from the file specified as [target file] that contain the string specified in [string]. The meaning of "SCF (self-consistent field) calculation" in the section title will be explained later, but that is "repeating the calculation until the evaluated energy value converges to a certain value" (called **iteration**). The calculated value in each stage is written out as "total energy = XXX" in "scf.out", so if you extract and display only the line containing the string "total energy", you will see the output shown above. You see that the energy values are rather oscillatory at first, but as you go through the steps, they become closer and closer to a constant value with about 4 decimal places of agreement. Thus, the quickly check whether the **SCF calculation is successful** can be performed by using "grep" to capture the energy values from the standard output and check for convergence.

Figure 3.4 shows how the energies behave as iteration proceeds. The energy values are converging as the iteration steps increases, and finally becomes "consistent" in the sense that the one at the k-step is almost unchanged at the $(k + 1)$-step. This process is described as 'SCF loop is running', and when the convergence has been achieved, it is described as 'SCF has successfully been converged', which corresponds to the 'dough being matured'.

It is commonly found such mistake by beginner that they forget to check the convergence. When they whinge "I followed the procedure to get DOS (described in the next section), but the result was totally strange...", the immediate thing to do is to execute "grep" as described above. Most of the case, you may find that the energy has not yet converged. Since the input for this tutorial has been prepared by the teacher, the convergence is easy as shown in Fig. 3.4. In the practical cases, however, the convergence sometimes required several hundreds iterations being beyond the default setting, 100. In such a case, the calculation once stops after 100 iterations, by which the convergence has not yet been achieved (in this case further resuming calculations required, as explained later in Sect. 10.2.3). Beginners, however, tend to misunderstand that the calculation has been completed, because all the calculations are actually finished without any error message with "save directory" generated. By using that "save directory" (not well-fermented dough), further analysis (e.g., DOS) are feasible but to get strange results.

> **Important**

Proficient users can find out the cause when consulted by a beginner under the condition that **"scf.out" is surely saved as a log**. If a beginner says "I've already erased it...", it would be impossible to find out the cause of the problem, so I would like to point out again the importance of saving the standard output file as a valuable log.

3.4   Quick Check Using Plotter

In this section, we will learn how to make a plot like Fig. 3.4 to check the results. This may seem like a trivial technical matter that should be located in an appendix, but we have included it as the main text because the author regards the skill described here is a quite important factor to keep one's motivation to learn, not to stall.

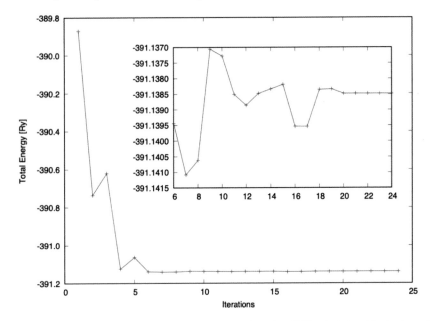

Fig. 3.4 Convergence of energy values (vertical axis values) in a SCF calculation. The horizontal axis is the number of iterations. The inset is a magnified view of the convergence from the sixth iteration. If the iterations are not sufficiently converged, then, in the metaphor used at the beginning of this chapter, it means that the "bread dough" has not yet matured, and even if you proceed to the later analysis to evaluate properties using it, you will get meaningless results. In an example-level case as shown in this figure, convergence is easily achieved after about 25 iterations. In some practical-level cases, the convergence may not be so easy, requiring several hundred iterations. The default input setting makes the calculation terminate when the iteration is repeated 100 times, and may be misunderstood by beginners as if the calculation has been successfully completed (because it terminates with some finite energy value!)

3.4.1 Useful Commands for Text Editing

If you can create a spreadsheet file like

```
1 -389.91622701
2 -390.74166474
3 -390.62950835
4 -391.13131932
...
```

you can imagine that a tool (plotter) can create a plot like Fig. 3.4 with the first column (number of iterations) on the x-axis and the second column (energy value) on the y-axis. Thus, the question is how to efficiently create the spreadsheet above from the output of "% grep 'total energy' scf.out". If you don't have the skills, you might start doing things like "cut and paste the numbers one by one with the mouse and then port them to Excel. The command line offers a full range of ways to process the data all at once without this kind of tedious "mouse-moving" tasks.

> **Important**

The important thing to remember is that "with a little effort, you can gain significant efficiency later on". This is where many people say, "I know, but...". Do spare time to memorize the command usage. Such metaphors come to mind taking medical checkups as example that "avoiding small, one-time expenses will result in many times more expenses over the long term". Touch-typing would be another good example. The author had actually avoided to memorize the usage of useful text processing command such as "grep", "sed", and "awk" (the topics we will learn in this section) for about 10 years, but instead maintained a handy notebook that summarized the usage. Every time I faced a task, I had to find where is the page on the note, and I would end up hitting the "space/delete/arrow-keys" repeatedly to move the cursor, thinking, "Oh well. that's faster than to find the note". If you do this, you will never be able to learn. Once I made up my mind that I can memorize, it was actually easily possible. Then I regretted that I should have memorized things earlier.

If you just did the command "% grep 'total energy' scf.out" in the previous subsection, **use the up-arrow key to reproduce the commands**, and then amend it with pipe (Sect. 9.2.2) as

```
% pwd
    /home/student/work/01tryout
% grep 'total energy' scf.out | awk '{print $4}'
    -389.09239119
    -390.91105457
    ...
```

A new command **awk** appears [4] in the form "awk '{processing content}'". Let's try changing the "$4" to $3 or $1 and execute it (use up-arrow key to amend). Then you will see that "awk '{print $4}'" means "extract only the information in the fourth column". While "**grep**" is for extracting lines, "awk" is for extracting columns, so you can imagine how these can be used to efficiently extract the "desired information" from a text file.

If you compare the output of

```
% grep 'total energy' scf.out | awk '{print $4}'
```

with the output of

```
% grep 'total energy' scf.out
```

you will notice that the last line

```
!    total energy              =    -391.13886725 Ry
```

is missing. This line starts with "!",[17] which is extra compared with other lines. As such, the value "−391.1..." is located in the 5th column instead of the 4th column, not capable to be captured by "awk '{print $4}'".

This problem can be solved by removing the "!" from the result of "grep 'total energy' scf.out". Try using the pipe as

```
%grep 'total energy' scf.out | sed 's/!//g'
```

The command "**sed**" appears with "'s/!//g'" as an argument. This argument is not easy to understand for beginners and looks like "sutra", leading sometimes to mistyping. This part should be read as "[action]/[string01]/[string02]/[applied range]". In this case, [action] is specified as "s" (substitution), for which [string01] is replaced by [string02] for the range "g" (global). As such, the above "sed" replaces '!' into '' (blank), being equivalent to remove appeared "!".[18] By executing the above, you see that "!" surely disappears. Then, with the pipe,

```
%grep 'total energy' scf.out |sed 's/!//g' |awk '{print $4}'
```

can extract all the energy values including the last line with the final converging energy.

---

[17] This is a marker for the final value to be converged, so that you can extract only the converged value by "grep ! scf.out".

[18] Type it with saying out "sed, substitute (s) exclamation mark (!) into black with global (g)", making you easier to memorize the usage.

To get the spreadsheet, we want the first column "1, 2, ..." corresponding to the *x* value. It can be generated by replacing the argument of "awk", "{print \$4}" into "{print NR, \$4}", as

```
% grep 'total energy' scf.out | sed 's/!//g' \
  | awk '{print NR, $4}'
```

Here, "NR" is the variable used in "awk", to which the line number is written. As such, the above command writes out the line number as the first column, and the energy value as the second column. Then,

```
% grep 'total energy' scf.out | sed 's/!//g' \
  |awk '{print NR, $4}' > temp
```

writes out the desired spreadsheet as the file "temp" using the redirect[19] :

```
% cat temp
    1 -389.91622701
    2 -390.74166474
    3 -390.62950835
    4 -391.13131932
    . . .
```

You will find that the "temp" includes the unwilling string "16 is" at the last line. Edit the "temp" and remove this line.[20]

Again, let's consider the comparison how to compose the spreadsheet using Excel with your mouse or by the commands using pipe and redirect. Once you've memorized the usage of the commands, the sheet is generated at a moment. No need to memorize everything, but only the usage. The location such as "\$4" or "\$3" depends on the situation, you can adjust it through trial and error. If your PC is connected to the Internet, you can use Google to find the usage of "sed", "awk", and "grep". Adding to these three tools, "gnuplot" as explained later have formed long history to be used to develop larger system infrastructures, providing a well-matured technology with large number of users all over the world. These technique is called "**text processing**" [4].

3.4.2   Script to Utilize of Past Knowledge

Suppose a situation where we have to repeat the above process with "scf.out" every time. Since the format of scf.out is always the same, it is inefficient to explore the

---

[19] The backslash "\" is used to show a too long string to fit on a line (see Acronym).

[20] Do not hit arrow keys to move the cursor. Do make your fingers memorize to get habit of using "Esc (and release once and) Shift+>" (moving to the last line), "Ctrl+A" (moving to the head of a line), and "Ctrl+K" (removing the part in a line after the cursor position).

settings ("$4" or "$3", etc.) for "sed" and "awk" by trial and error for each time. Once we have figured out the proper setting with a little effort as shown below, we can use the same for the **next time**. Showing the previous command (as shown below) on the terminal by using the up-arrow key, capturing it by "cut and paste", and then open a new file as "emacs script01". Edit the file so that its contents can be[21]

```
% pwd
    /home/student/work/01tryout
% emacs -nw script01
    [After completing required edit, close the file]
% cat script01
    grep 'total energy' scf.out | sed 's/!//g' \
    |awk '{print NR, $4}' > temp
```

Note that the cat command is not intended for you to show the contents of the file. It is rather to instruct the user to check if the contents you have edited are surely look like this. It is customary to indicate "please edit the contents of your file like this" by showing the output of the cat command in this way.

At this stage, "script01" is "just a file", but we can use this file as a "new command". To promote it to a "command", execute

```
% chmod u+x script01
```

Here, the command **chmod** has its mnemonic as "change mode".[22] The command works as "change the mode of 'script01' into 'u+x'". To understand what is "u+x", let's introduce the concept of **file permission**. Each file has its attribute as the combination of "readable (r)"⊗"writable (w)"⊗"executable (x)". Each property can be switched to "enable (+)" or "unable (-)" for "the owner (u)" or "the group member (g)" or "the other (o)". When we say "promote a file to a command", what we really mean is "make the file attribute executable". The "u+x" is then read as "enabling the executable attribute for the file owner", corresponding to that the owner can execute the file as a command. You can check if the "chmod" command has correctly changed the attributes by using the "% ls -l" to display the file list.

! Attention

Search Google by yourself how you can examine the file attributes from its displaying list.

To execute a "command created from a file" in this way, you cannot do it with "% script01", but you need to specify the path to the file name explicitly. Since "script01"

---

[21] The backslash "\" is used to show a too long string to fit on a line (see Acronym).
[22] The fastest way to memorize many commands is to always say its mnemonic out loud during you type the command.

is located in your current working directory, you can run it as (remember what "./"
means)

```
% ./script01
```

Execute the above after you remove the expected output by the "script01" as "% rm
temp". By executing the above, you can confirm that new "temp" is created by using
"ls". The "temp", however, includes the improper line "16 is" as explained in the
previous subsection.

    If you search by Google for "Linux, deleting the last line", you can get a lot of
information how to delete only the last line of a file by any command. One way to
do this is to use "sed '$d' temp" to delete the last line of "temp".[23] Then, execute
"cp script01 script02" to make a copy, and edit it as[24]

```
% cat script02
   grep 'total energy' scf.out | sed 's/!//g' \
     |awk '{print NR, $4}' > temp
   sed '$d' temp > temp2
```

Remove "temp" in advance (% rm temp), and then execute

```
% ./script02
```

You can confirm that the execution creates "temp" as well as "temp2". When you
"cat" the "temp2", you will find that it does not include the unwilling line "16 is".

    In this way, you can record a set of multiple commands together on a file as
"one command" and execute it. This is called **script work**. As such, you can now
understand how it works:

```
% cat script03
   grep 'total energy' scf.out | sed 's/!//g' \
     |awk '{print NR, $4}' > temp
   sed '$d' temp > temp2
   rm temp
   mv temp2 temp
```

Edit the above starting from a copy of script02.
    Before executing "script03", try

```
% rm temp*
```

---

[23] The basic usage is "sed '10d' temp" to delete the 10th line. By replacing "10" into "$", the "$"
works as the variable to represent the last line.
[24] The backslash "\" is used to show a too long string to fit on a line (see Acronym).

Here, "*" is called "wild card", which stands for everything. Namely, "temp*" means "any name starting from 'temp' followed by any string". The "rm temp*" can then remove temp as well as temp2 at once. The wild card is one of the powerful functionalities in Linux operation for massive processing. After the deletion, execute

```
% ./script03
```

and confirm the final "temp" is made as what you want.

In this way, the script work can be conveniently used to perform automatic and mass processing with flexibility. In Sect. 3.3.1, we mentioned that "it is a good idea to use the same name for input and output files". This is because it is easier to compose a script assuming the same file name written as its contents.[25]

3.4.3   Using a Plotter

Up to the previous subsection, we described how to create a data file to be processed by the plotter. Next, we will learn how to use the plotter. For the plotter, we will use "gnuplot" application, which can be easily installed on any Linux system and is free. It is a highly versatile plotter with many users all over the world. It can create graphs to be inserted into papers at the proficient level.

> **! Attention**
>
> The operations in this section should be done by logging directly into the RaspberryPi, not by "remotely connecting from a PC" as there are "on-screen plotter illustrations".

When you run the command "gnuplot" from a terminal, the prompt "gnuplot>" appears waiting for your input as

```
% pwd
    /home/student/work/01tryout
% gnuplot
    G N U P L O T
    Version 5.4 patchlevel 1    last modified 2020-12-01
        ...
    gnuplot>
```

Try executing "quit" for the prompt, then the gnuplot application is quitted, getting the normal terminal prompt back.

Then, try

---

[25] It is, of course, possible to elaborate by automatically extracting the file name and assigning it to a variable, but it is better to keep things as simple as possible so that one can easily understand how the script works when one reads it.

```
% pwd
    /home/student/work/01tryout
% ls temp
    temp
% gnuplot
    G N U P L O T
    Version 5.4 patchlevel 1     last modified 2020-12-01
        ...
    gnuplot> plot 'temp'
```

In the gnuplot deck, you can use the up- and down-arrow keys to reproduce the previous commands, and tab complete as well. Make your typing more efficient with these functions. By the "plot" command, you will see that a plot appears like Fig. 3.4. To exit from gnuplot, just type on the deck as "q" instead of quit.

```
        ...
    gnuplot> q
```

In gnuplot, this kind of single-letter abbreviation often works.

Let's start up gnuplot again and follow the sequence below[26]:

```
gnuplot> p 'temp'
gnuplot> p 'temp' pointsize 3
gnuplot> p 'temp' pointsize 3 pointtype 4
gnuplot> p 'temp' ps 3 pt 4
gnuplot> p 'temp' ps 3 pt 4, 'temp' with line
gnuplot> p 'temp' ps 3 pt 4, 'temp' w l
```

In each line, "p" is a single-letter abbreviation of the command "plot". The "pointsize 3" changes the size of the symbol, and you can try changing the "3" to various values to see how the size changes. The "pointtype 4" changes the default symbol shape. You can also change the value here to see how it works. You can abbreviate pointsize and pointtype to "ps" and "pt", respectively. When typing "ps", make a habit of saying "pointsize" as you type to help you remember faster. If you add "..., 'temp' with line", it will show the plotted points connected by a broken line. The "with line" is abbreviated as "w l".

As you will learn, gnuplot also allows you to add labels to the x- and y-axes, to move the legend position, to change the color of the plot, etc. In order to save time and effort to set the symbol size, color, etc., you can create a template of such settings as a script in advance and load it into the plot (explained later in Sect. 3.5.2). In combination with these functions, it is possible to program a series of scripts to automatically create a plot by a single command after the simulation is completed.

---

[26] Don't forget to use the up-arrow to reuse the previous command, and Ctrl+A, Ctrl+E, and Ctrl+K.

3.5   Calculating Electronic Structure

In line with the metaphor of "dough" used at the beginning of this chapter, the content of the previous section corresponds to "making bread dough that has matured sufficiently. This section is about baking tin-bread (Fig. 3.1) from the dough to see if the dough is firm enough.

In the previous section, we ran the SCF loop to get the converged self-consistent electronic state (wave function). The calculations of this section use this converged result. So we no longer need to run further SCF loop. We will refer to these calculations as "non-SCF" and label them as **NSCF calculations**.

It is one of the typical mistakes made by beginners who do not understand the above to run the new SCF calculation from the beginning again to get what NSCF provides. For NSCF, there is a way to prepare the input file so that the calculation can be performed using the **existing** converged wave function prepared by the previous calculation. Even with the mistake, the result will be the same, but note that SCF sometimes takes a couple of days while a shot of NSCF takes several seconds. The mistake severely wastes the computational resource.[27]

In this section, we will learn the procedures how to get DOS (density of state) shown as total and partial DOS (pDOS), as well as the band dispersion diagram as "the calculation of electronic structure". The DOS and dispersion diagrams [1] are used in the same way that a medical doctor uses an X-ray picture to consider the state of a patient. They provide the most fundamental information for researchers to consider the possible scenarios. However, for the time being in this course, we will regard them just as equivalent to "fingerprints" of the electronic state of the system. Actually, it is sometimes used to check whether your calculations are reasonable or not by first drawing them and then comparing them with dispersion diagrams reported in previous studies.

In solid-state physics, the order of learning is "band dispersion first to get DOS" [1]. However, in practice of simulations, DOS can be depicted with much simpler operations than those required for dispersions. So, usually, users get DOS first, and if necessary, some additional operations (editing the input file with a little more complicated additions) will be performed to get the dispersion. This course also follows this procedure.

> **Important**

The band dispersion is a typical unique concept appearing in the theory of solid periodic systems, where one considers things in the wave vector space. When I was conducting joint research with industry, a young researcher with a background in molecular science asked me, "To be honest, I haven't really understood for a long

---

[27] In fact, there are some beginners who are occupying the resource with such useless calculations on shared computers in this way. Note that the most of proficient users can identify such useless ones from their job names with who is running that [5]. Be careful not to be spotted by tougher seniors...

time, but why are band dispersion always presented? What should I read from it?". This was a very good subject for teachers to answer, for which a possible systematic and concise explanations are given in Chap. 11.

For both DOS and dispersion calculations, each NSCF calculation starts off from the same SCF converged result obtained in the previous section. Therefore, these calculations are performed on the same working directory where the SCF calculation was performed.

3.5.1 NSCF Calculations for DOS

We prepare the input file for the NSCF calculation (nscf_dos.in) for DOS by amending the input for the SCF calculation used in the preceding section.

> **! Attention**
>
> From here, you can go back to connecting remotely from your familiar/comfortable PC.

```
% pwd
   /home/student/work/01tryout
% cp scf.in nscf_dos.in
```

Since "scf.in" is an important research log for the SCF calculation, we do not edit it directly, but use "cp" to copy it and then amend it. Open the "nscf_dos.in" using emacs and edit it as follows[28] :

```
&control
calculation = 'nscf' ! modified here (1)
...
outdir='out_qa'
...
/
&system
occupations = 'tetrahedra' ! modified here (2)
...
/
```

---

[28] To avoid typos, the input files used in this course are included as text files in the downloaded materials as introduced in Sect. 2.4.2. You can copy them, or "cut and paste", or for comparing to find mistakes using "diff" or "sdiff" command. These files are stored with the same file name under "~/work/setupMaezono/inputFiles/".

```
K_POINTS {automatic}
4 4 4 1 1 1 ! modified here (3)
...
```

The lines with the comment, "! modified here", should be modified as shown above. In "!.... (1)" to change the calculation mode from SCF to NSCF. (2) and (3) will be explained in detail in the next chapter, but this part specifies how fine the numerical integral mesh, and the modification is made to increase the accuracy since the physical properties are evaluated here. In the previous SCF, the calculation is to achieve the convergence as soon as possible with the minimum mesh accuracy. The NSCF calculation is a "one-shot calculation after convergence", so it just takes the CPU time cost for one loop in SCF. In general, higher mesh accuracy is preferable for the evaluation of physical quantities, so we set the mesh accuracy to a certain level higher than the one used for the SCF calculation.

Prepared the input file as described above, execute the NSCF calculation for DOS using "redirect" as in the case of SCF calculation[29],[30]:

```
% pw.x < nscf_dos.in | tee nscf_dos.out
```

3.5.2 Depicting DOS

Completing the NSCF calculation by "nscf_dos.in", we prepare another input file for the DOS depicting calculation as "dos.in" on the same working directory as[31]:

```
% pwd
    /home/student/work/01tryout
% emacs -nw dos.in
    [Edit 'dos.in' as shown as the output of 'cat' command below]
% cat dos.in
    &dos
    outdir = './out_qa',
    prefix='qa',
    fildos='qa.dos',
    /
```

While "nscf_dos.in" is used to run "pw.x" to **generate the wave function** to evaluate DOS, the "dos.in" is used to run another binary "dos.x", which is used to **evaluate**

---

[29] A bug has been observed in some versions of Quantum Espresso that causes this calculation to stop. In this case, adding the line "nosym=.true." to the "&system" entry in "nscf_dos.in" will solve the problem.

[30] It will take around 20 min.

[31] To avoid typos, the input files used in this course are included as text files in the downloaded materials as introduced in Sect. 2.4.2. You can copy them, or "cut and paste", or for comparing to find mistakes using "diff" or "sdiff" command. These files are stored with the same file name under "~/work/setupMaezono/inputFiles/".

DOS from a given wave function. In "dos.in", it is specified that the output of the calculation is dumped to "./out_qa" (the same as the existing save directory used by the previous SCF calculation), and the DOS is written in the file named "qa.dos".

The binary, "dos.x", is provided in the Quantum Espresso package, located at the same hierarchy as the binary "pw.x" that you have been using so far. Run it as follows:

```
% dos.x < dos.in | tee dos.out
```

The calculation takes only a few seconds. When it is over, you can see that two files, "qa.dos" and "dos.out", have been created. While "dos.out" is the standard output file for the calculation, another output, "qa.dos", contains the numerical data required for drawing DOS, as shown below.

```
% cat qa.dos
#   E (eV)     dos(E)      Int dos(E) EFermi =     6.927 eV
 -16.844   0.0000E+00   0.0000E+00
 -16.834   0.7721E+00   0.7376E-02
 -16.824   0.8280E+00   0.1538E-01
 -16.814   0.8844E+00   0.2394E-01
 ...
```

The first column is the energy value that will be the horizontal axis of the plot, and the second column is the value of the density of states used as the vertical axis of the plot. The third column is the integrated density of states value (corresponding to how many states exist below that energy value).

If you have already mastered Sect. 3.4, you might want to plot the above spreadsheet using "gnuplot". In Sect. 3.4, we used "gnuplot > plot 'temp'" for a spreadsheet "temp" that contains two columns of data. Even when "temp" contains three or more columns of data, we can still use "gnuplot >textgreater plot 'temp' using 1:2" to plot the first column on the x-axis and the second column on the y-axis. Remembering that "plot, and "using" can be abbreviated by one letter, you can choose "which column to plot", for example, "gnuplot > p 'temp' u 1:3" would plot the third column on the y-axis.

! Attention

Since the operations from now include the plot output on the screen, they should be directly on the RaspberryPi. If you are operating it remotely from your familiar PC (Sect. 2.3.3), please switch to login directly to the RaspberryPi now (it is reminded later at the point where you can get back to the remote connection).

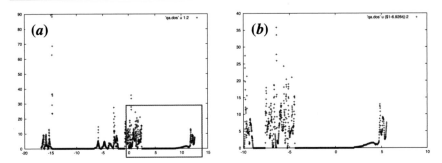

Fig. 3.5 Density of states (DOS) for α-quartz. When the data file is plotted as it is, an overall view like (**a**) appears. By adjusting the display range by shifting the zero of the horizontal axis to the Fermi level, the red box in (**a**) is enlarged, resulting in the plot shown in (**b**)

By executing

```
% pwd
    /Users/maezono/work/01tryout
% gnuplot
    gnuplot> p 'qa.dos' u 1:2
```

you will get the plot as shown in Fig. 3.5a. As expected, all the ranges contained in dos.out will be depicted. What is interesting in the DOS, however, is the gap presence/absence and shape near the Fermi level. From Fig. 3.5a, however, it is not clear where the Fermi level is. The Fermi energy is given in the "nscf_dos.out" of the NSCF calculation, and can be captured by grep as follows:

```
% grep Fermi nscf_dos.out
    the Fermi energy is    6.9266 ev
```

By subtracting this value from the x-axis value, we can get a "plot with the Fermi level as zero" as

```
gnuplot> p 'qa.dos' u ($1-6.9266):2
gnuplot> set xrange [-10:10]
gnuplot> p 'qa.dos' u ($1-6.9266):2
```

Do the above line by line, to get the plot as you want.[32]

Various options at the gnuplot prompt appear in the above, which are explained below: First, "u ($1-6.9266):2" specifies that 6.9266 eV will be subtracted from the numerical value of the first column data (adopted for the x-axis). The word "$1" means "the variable in which the first column data is stored". The "set xrange [-10:10]" specifies the display range of the x-axis. Then, by plotting the data again

---

[32] Use the up-arrow key to display the previously executed commands and make corrections.

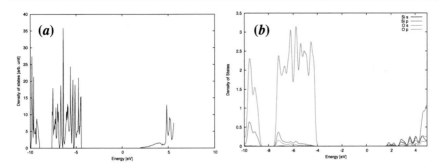

Fig. 3.6 Density of states (DOS) [panel (**a**)] and partial DOS (pDOS) [panel (**b**)] evaluated for α-quartz

(the third line), we get a plot like Fig. 3.5b. From this figure, we can identify that the silicon dioxide is predicted to be an insulator because of the zero DOS just above $x = 0$ (Fermi level) (toward the right along the x-axis).

Quitting gnuplot (by typing "q" at the gnuplot prompt), and then start the gnuplot again. Then, execute line by line as follows:

```
gnuplot> set xlabel 'Energy [eV]'
gnuplot> set ylabel 'Density of states [arb. unit]'
gnuplot> set xrange [-10:10]
gnuplot> p 'qa.dos' u ($1-6.9266):2 w l notitle
```

This should give you a plot like Fig. 3.6a. The "set x(y)label" specifies the heading of the $x(y)$-axis as specified by the string enclosed by the single quotations. "w l" is an abbreviation for "with lines", which displays the data as a line. The "notitle" option specifies not to display the legend "'qa.dos' u ($1-6.9264):2" that was previously shown at the upper right corner. You can learn more about these options by searching Google for "gnuplot".

Once you have completed the above plot as you want, you don't need to type the same on the keyboard every time you need to do the same. Store the above operations in a text as

```
% cat dosPlot.gv
    set xlabel 'Energy [eV]'
    set ylabel 'Density of states [arb. unit]'
    set xrange [-10:10]
    p 'qa.dos' u ($1-6.9266):2 w l notitle
```

When you start up a new "gnuplot" session, execute

```
gnuplot> load 'dosPlot.gv'
```

Then the setting written in the gv file is loaded. This is also a kind of "script work" for gnuplot, by which you can accumulate your previous experiences as the knowledge.

3.5.3 Depicting Partial DOS

The partial density of states (pDOS/partial DOS) is the decomposed the DOS into each orbital component (s, p, d, ...) to identify each contribution. By using pDOS, we can argue which orbital components contribute to the properties especially dominated by the vicinity of the Fermi level, or which components to the lowest energy excitations. While the utility "dos.x"[33] was used to depict DOS (total DOS) in the previous subsection, we use "projwfc.x"[34] for pDOS.

Prepare the input file for "pdos.in" as[35]

```
% pwd
    /home/student/work/01tryout
% emacs -nw pdos.in
    [Edit 'pdos.in' as shown as the output of 'cat' command below]
% cat pdos.in
    &projwfc
    outdir = './out_qa',
    prefix='qa',
    degauss = 0.01
    /
```

As summarized later in Fig. 3.7, please note the dependency that both "dos.x" and "projwfc.x" are processing the same result of the SCF calculation by pw.x, located in the same directory, "out_qa". The "out_qa" was specified in the input file for "pw.x" (Sect. 3.3.1). In response to this, the above input "pdos.in" for "projwfc.x" specifies "outdir = '. /out_qa'".

Then, execute "projwfc.x" as

```
% projwfc.x < pdos.in | tee pdos.out
```

After the execution, you can confirm that it generates the following files as its output:

```
% pwd
    /home/student/work/01tryout
% ls
    qa.pdos_atm#1(Si)_wfc#1(s)    qa.pdos_atm#1(Si)_wfc#2(p)
    qa.pdos_atm#2(Si)_wfc#1(s)    qa.pdos_atm#2(Si)_wfc#2(p)
    qa.pdos_atm#3(Si)_wfc#1(s)    qa.pdos_atm#3(Si)_wfc#2(p)
    qa.pdos_atm#4(O)_wfc#1(s)     qa.pdos_atm#4(O)_wfc#2(p)
    ...
    qa.pdos_atm#9(O)_wfc#1(s)     qa.pdos_atm#9(O)_wfc#2(p)
```

---

[33] While "pw.x" solves the equation to get wave functions as the primary role in the package, "dos.x", etc. take the secondary role to analyze the results, those are called "utilities".

[34] A mnemonic that combines "projecting" onto each component with the "wavefunction".

[35] To avoid typos, the input files used in this course are included as text files in the downloaded materials as introduced in Sect. 2.4.2. You can copy them, or "cut and paste", or for comparing to find mistakes using "diff" or "sdiff" command. These files are stored with the same file name under "~/work/setupMaezono/inputFiles/".

The filename ended with (s) and (p) includes the pDOS for s- and p-orbital components. You notice that even for the same element and the same component, there are
different files as "...#1(Si)..._(s)", "#2(Si)..._(s)", etc., being corresponding to each
different atomic site within a unit cell. In the present case, there are three sites for Si
and six for Oxygen, but they are equivalent leading to the contents is the same regardless of "#x". Looking into the contents of each file, you can see data spreadsheet as
in the form of "qa.dos".

To depict the data, open "gnuplot" and execute[36]

```
set xrange [-10:5]
set xlabel 'Energy [eV]'
set ylabel 'Density of States'
plot\
'qa.pdos_atm#1(Si)_wfc#1(s)' u ($1-6.9266):2      w l \
    title 'Si s', \
'qa.pdos_atm#1(Si)_wfc#2(p)' u ($1-6.9266):2      w l \
    title 'Si p',\
'qa.pdos_atm#4(O)_wfc#1(s)'  u ($1-6.9266):($2*2) w l \
    title ' O s',\
'qa.pdos_atm#4(O)_wfc#2(p)'  u ($1-6.9266):($2*2) w l \
    title ' O p'
```

Execute them line by line, considering each meaning and not forgetting to use arrow
keys and tab completion. By using "plot \", the prompt behaves as

```
gnuplot> plot\
>
```

Then, you can type

```
''qa.pdos\_atm\#1(Si)\_wfc\#1(s)''...
```

The form

```
  plot [the first process], [the second process], ...
```

works for the multiple plots to be appeared on the same panel. It would lead to too
long contents in a line, for which we can avoid it by using "\" as in the above. Adding
"title '[string]'" works to set the **legend label** for each data. After these settings, you
will get the plot as shown in Fig. 3.6.

For oxygen plot, you may have noticed multiplier of 2 as in "($2*2)". This is
because the number of atoms in the unit cell is "3 for Si" and "6 for O", so the weight
of each file is 1/3 and 1/6, respectively. By displaying Si and O in the ratio of 1:2,
the display will correspond to the meaningful weights.

---

[36] The backslash "\" is used to show a too long string to fit on a line (see Acronym).

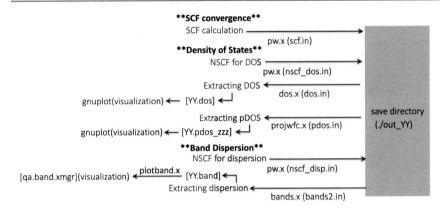

Fig. 3.7 The workflow for depicting DOS and band dispersion diagrams. The ingoing arrow to the save directory means that the calculation dumps the output in the directory. The outgoing arrow from the directory means that the contents of the directory are referred by the calculation. "**.x (**.in)" in red indicates the binary and input files used for processing. "YY" is the label of the system (it is "qa" in this chapter), and "zzz" is the designation of the elemental and orbital species. [Filename] means an extra output file in addition to the standard output

The procedure flow explained so far is summarized in Fig. 3.7. Please pay attention to what input file you used and what binary you called as shown in the figure.

> **! Attention**
>
> From here, you can go back to connecting remotely from your familiar/comfortable PC.

3.5.4 Depicting DOS Using xFroggie

In the above, we have explained DOS drawing using gnuplot. In these explanations, one had the opportunity to learn the meaning of the values in the output file through various command operations and text processing realizing the shifting of the origin value, multiplying of the y-value, etc. The drawing using gnuplot has the above educational meaning. On the other hand, for practical use, you can use the convenient web application "xFroggie" to quickly check DOS, as described in this section.

Visiting the website,[37]

```
https://xfroggie.com
```

click the hypertext "Plot DOS and PDOS obtained by QE".

---

[37] The referenced URLs may have changed, so please refer to the latest link information given in the text file ~/work/setupMaezono/memo.txt included in the downloaded set of materials.

Plot DOS and PDOS obtained by QE

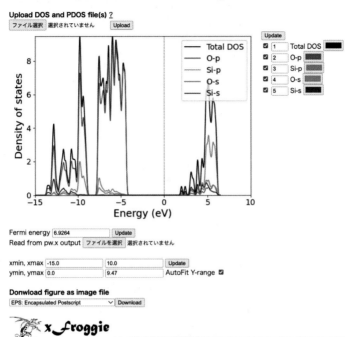

Fig. 3.8   DOS and pDOS depicted by "xFroggie"

Press the "Choose File" button beneath the text "Upload DOS and PDOS file(s)", and select all the pDOS files ["qa.pdos_atm#1(Si)_wfc#1(s)", etc.] as well as the total DOS file [qa.pdos_tot] at once.[38] Then, press "Upload" button to upload them. DOS will be depicted soon on the browser.

Beneath the depicted DOS, one can find the text "Fermi energy" and the "Choose File" button below the text. Clicking the button to select "nscf_dos.out" and upload it. Press the "Update" button, the DOS is depicted again with the modification so that its $x = 0$ corresponds to the Fermi level.

You can further adjust the drawing area by entering values in the "xmin/xmax/ymin/ymax" fields and clicking the "Update" button. Also, if "AutoFit Y-range" is checked, the appropriate y display range for a given x display range will be automatically taken.

After the DOS is depicted, a list of legends is displayed on its right side as shown in Fig. 3.8. You can specify whether or not to display the items by the checkboxes in the list, the order of the legend from the number input boxes, and the line color from the color boxes. These specifications are reflected when DOS is depicted again

---

[38] To select multiple files at once, hold down some key [Shift or Cmd or Ctrl depending on your OS (just try!)] and click on multiple target files.

by clicking "Update" button. For example, to move "Total DOS" to the top of the legend, enter "1" in the numeric entry field and click the "Update" button, then the order of the other legends will automatically move down by 1.

Finally, you can download the DOS diagram in any formats, "eps/jpeg/pdf/png/ps/tif/..." on your selection by clicking on "Download figure as image file" at the bottom of the page.

3.6   Depicting Band Dispersion

As mentioned in Sect. 3.5, band dispersion plotting requires a bit more detailed specification than DOS, requiring the user to predetermine (1) the number of bands to be drawn, N_{bnd}, and (2) the k-point path. N_{bnd} includes the valence bands (occupied bands) as well as the conduction band (unoccupied ones), by which you specify up to which unoccupied level the diagram is shown (note that, in principle, there are infinite bands toward high energy).

The k-point path is introduced to depict the $\mathbf{k}$-dependence of the energy band, $\{\varepsilon_b(\mathbf{k})\}_{b=1}^{N_{bnd}}$. The dependence means that there are N_{bnd} values for a fixed point, $\mathbf{k}$, in the three-dimensional space of (k_x, k_y, k_z). The N_{bnd} values vary as the position of $\mathbf{k}$ moves, providing N_{bnd}-fold curves, called dispersion.

Since $\mathbf{k}$ is three dimensional, the values are given for each point in three-dimensional space. To depict the behavior shown on two-dimensional figure, the k-point path is used. Taking a typical path connecting characteristic points in three-dimensional $\mathbf{k}$ space, as shown in Fig. 3.9, the values, $\{\varepsilon_b(\mathbf{k})\}_{b=1}^{N_{bnd}}$, would vary along the path (k-point path). This behavior is shown as the band dispersion diagram.

Fig. 3.9 A k-point path set in the Brillouin zone of a FCC (face-centered cubic) lattice. There are several characteristic points like Γ, X, L, etc. in the three-dimensional space of (k_x, k_y, k_z). There is the convention to take a path connecting these points. The dispersion diagram is shown as the change of quantities depending along the path

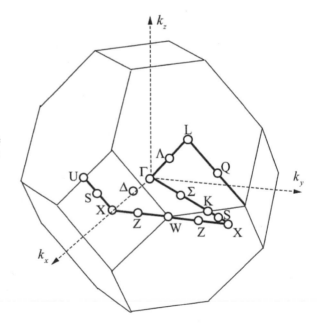

For each crystal symmetry, there is a convention how the k-point path is taken. On the same convention, you can compare your result with other reported ones. Since the conventions get to be integrated as time going, sometimes you would face to the different looking illustrations based on the different conventions in the old literatures. In the past, we had to look at the reference providing the k-point path convention (typically at an appendix of a textbook), picking up its coordinate and editing the input file by hand. Nowadays, however, there are many useful tools that can automatically generate the k-point path written out in a file. In Sect. 10.1, the procedure how to generate the k-point path using a web tool "xFroggie" is explained.

For given N_{bnd} and k paths, the procedure to depict the dispersion diagram is as follows (Fig. 3.7):

- (a) performing a NSCF calculation for band dispersion ("pw.x"),
- (b) extracting required information for the diagram ("band.x"),
- (c) depicting the plot ("plotband.x"),

and we will explain them.

3.6.1 NSCF Calculation for Band Dispersion

N_{bnd} is determined by the following procedure. The "number of electrons in the unit cell" is written out in the output file "scf.out" of the SCF calculation, and it can be captured as

```
% pwd
    /home/student/work/01tryout
% grep ''number of electrons' scf.out
        number of electrons  =  48.00
```

The band index "b" in $\{\varepsilon_b(\mathbf{k})\}_{b=1}^{N_{bnd}}$ is like remnants of atomic levels, and hence a band can accommodate two antiparallel pairs of spins, $\uparrow\downarrow$ from lower level. This means that two electrons per unit cell occupy one band. In the present case, since there are 48 electrons in the unit cell, the 24th band is the upper occupied band.

For the conduction band (unoccupied band), about half the number of occupied bands would be enough for a normal analysis,[39] so let's set $N_{bnd} = 36$, including "24 for valence bands" + "12 for conduction bands".

---

[39] No problem to increase or decrease slightly from this decision. Note, however, that the larger the choice, the more computational cost will be required to identify all of the eigenvalues for the specified number of bands, so it is necessary to keep it at a reasonable level. If you double the number of bands, the computational cost will increase by a factor of 2^3 in most cases (Sect. 5.4.4).

Then, copy "nscf_dos.in" used for the DOS-NSCF calculation to "nscf_disp.in", and edit it as the input for dispersion-NSCF. First, amend the following[40]:

```
&control
...
calculation  = 'bands'
..
/
&system
...
occupations = 'fixed'
nbnd = 36
...
/
```

The "nbnd = 36" is the setting for N_{bnd}. In addition to the above, add the following set of "KPOINTS blocks" to the end of "nscf_disp.in":

```
K_POINTS {crystal_b}
12
0.0 0.0 0.0 20
0.5 0.0 0.0 20
0.3333333333333333 0.3333333333333333 0.0 20
0.0 0.0 0.0 20
0.0 0.0 0.5 20
0.5 0.0 0.5 20
0.3333333333333333 0.3333333333333333 0.5  20
0.0 0.0 0.5 20
0.5 0.0 0.5 20
0.5 0.0 0.0 20
0.3333333333333333 0.3333333333333333 0.0 20
0.3333333333333333 0.3333333333333333 0.5 20
```

This part is the "k-point path specification" explained above. The "12" appeared in the first line is the "number of k-points to be connected by the path", and the 1~3 columns in the following lines are the coordinates of each **k**-point. The "20" at the end of the line specifies that "20 points should be sampled between each k-point" (which determines the resolution of the diagram). The question is "how to get this k-point path", which will be explained in detail in Sect. 10.1. At this stage, we assume that we have given the path generated by the external tool.

Then, execute "pw.x" with the input "nscf_disp.in" (it will take around 4 hours by RaspberryPi):

---

[40] To avoid typos, the input files used in this course are included as text files in the downloaded materials as introduced in Sect. 2.4.2. You can copy them, or "cut and paste", or for comparing to find mistakes using "diff" or "sdiff" command. These files are stored with the same file name under "~/work/setupMaezono/inputFiles/".

```
% pwd
    /home/student/work/01tryout
% pw.x < nscf_disp.in | tee nscf_disp.out
```

Completing the calculation, the required information to depict the band dispersion is dumped to the save directory (Fig. 3.7).

3.6.2 Preparing Band Dispersion Data Using "bands.x"

By the calculation with "nscf_disp.in" in the previous subsection, the energy eigenvalues, $\{\varepsilon_b(\mathbf{k})\}_{b=1}^{N_{bnd}}$, for each of the $\mathbf{k}$-points were calculated and stored in the save directory. By using that fundamental quantities, the utility "bands.x" can create the band dispersion diagram (Fig. 3.7). The input file, "bands.in", for the "bands.x" is prepared as follows[41]:

```
% pwd
    /home/student/work/01tryout
% cat bands.in
    &bands
    outdir = './out_qa',
    prefix='qa',
    filband='qa.band',
    lsym=.true.
    /
```

The input specifies that it handles the contents in the target directory, "outdir = './out_qa'", and its output is written out to the file "qa.band". Then, execute the calculation as follows (it will take around 15 min):

```
% bands.x < bands.in | tee bands.out
```

Completed the calculation, you can find the three files, "qa.band", "qa.band.gnu", and "qa.band.rap" are written out in the working directory. These include essentially the same information, but in different format. We use "qa.band" to draw the band dispersion.

---

[41] To avoid typos, the input files used in this course are included as text files in the downloaded materials as introduced in Sect. 2.4.2. You can copy them, or "cut and paste", or for comparing to find mistakes using "diff" or "sdiff" command. These files are stored with the same file name under "~/work/setupMaezono/inputFiles/".

3.6.3   Depicting Band Dispersion Using "plotband.x"

As shown in Fig. 3.7, "qa.band" can be visualized by the utility "plotband.x". When executed, it produces two output written in different formats, xmgr and ps.[42] While the xmgr is used for full-fledged purposes, the ps format can be used for a quick check to look at the graphics.

To use "plotband.x", you need to specify [the display range in absolute values], [the value of the Fermi level E_F (obtained from the SCF calculation)], [the tick width of the vertical axis], etc. Referring to the DOS in Fig. 3.5, the display range should be around [0, 14], and correspondingly the tick being $\Delta \sim 1.0$. The E_F was 6.9266 eV as picked up from "nscf.out" in Sect. 3.5.2.

To reflect the above, edit the input file for "plotband.x" as follows[43] :

```
% cat plotband.in
    qa.band          !* Target filename to be processed
    0 14             !* Display range
    qa.band.xmgr     !* Filename of output in xmgr format
    qa.band.ps       !* Filename of output in ps format
    6.9266           !* Energy value of the Fermi level
    1 6.9266         !* Tick for vertical axis, Energy value of the Fermi level
```

By using this input file, execute "plotband.x" as

```
% plotband.x < plotband.in
```

Then, you can confirm that there are "qa.band.xmgr.XXX" (where XXX is a number) as well as "qa.band.ps".

> **! Attention**
>
> The following operation needs to be done directly on the RaspberryPi because of the handling of graphical pictures. If you are operating it remotely from your familiar PC (Sect. 2.3.3), please switch to login directly to the RaspberryPi now (it is reminded later at the point where you can get back to the remote connection).

---

[42] "xmgr" is a format that can be visualized using a plotter called "Grace", and "ps" is a postscript format.

[43] To avoid typos, the input files used in this course are included as text files in the downloaded materials as introduced in Sect. 2.4.2. You can copy them, or "cut and paste", or for comparing to find mistakes using "diff" or "sdiff" command. These files are stored with the same file name under "~/work/setupMaezono/inputFiles/".

Fig. 3.10 Depicted band dispersion from "qa.band.ps". It is just for a quick check of the shape without any indication as to what k-point path the horizontal axis corresponds to. For fully fledged picture including such indications can be created in xmgr format by editing the filling in

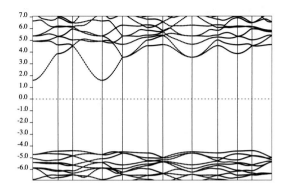

"qa.band.ps" is an image file as it is. To open it, we use the command "evince" as

```
% evince qa.band.ps
```

to get a picture as shown in Fig. 3.10.[44]

Then, type "Ctrl+C" to exit from "evince" (showing a picture).

3.6.4 Depicting Band Dispersion Using "xFroggie"

Also for depicting band dispersions, the web application "xFroggie" mentioned in Sect. 3.5.4 can be used for more quick check. Visiting the xFroggie website, click on the hyperlink "Plot band structure obtained by QE".

In the item stated "Upload band structure file (.gnu)", click the "Select file" button to upload "qa.band.gnu" and click "Update" to draw the dispersion diagram. At the bottom of the diagram, one can find the text "Fermi energy", and then upload "scf.out" and click "Update" to read the Fermi level. The figure will then be depicted again so that the Fermi level is set as $y = 0$. Enter values in the "ymin" and "ymax" fields directly below the Fermi energy specification and click "Update" to adjust the drawing area.

Under the text "High symmetry points", one can upload "bands.out" file. Then one will see the band dispersion as shown in Fig. 3.11. As shown in the figure, the correspondence between the x-axis value and the k-coordinate given in "nscf_disp.in" is shown on the right side.[45] One can specify the conventional label of high-symmetric points like Γ or L (Fig. 3.9) to be shown as in the figure.

---

[44] If the image is displayed vertically, click on the three-stripe icon in the upper right corner of the screen window, and you will see a selection for rotating the picture.

[45] The displayed k-point coordinate values are different from those in "nscf_disp.in", but this is because the former employs Cartesian coordinate values in reciprocal lattice space, while the latter employs coefficient values of reciprocal lattice vectors.

Plot band structure obtained by QE

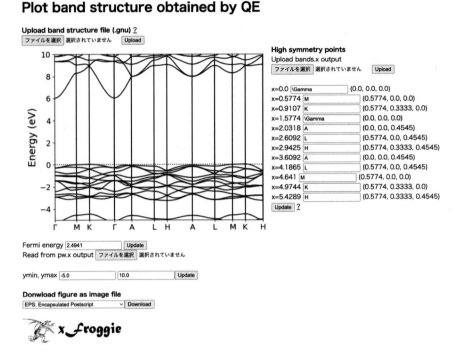

Fig. 3.11 Band dispersion depicted by "xFroggie"

As in Sect. 3.5.4, you can download the dispersion diagram as well as in any image format, executed at "Download figure as image file".

3.7   Calculating Properties as a Quick Check

In this chapter, we have explained how to depict the band dispersion and the DOS as a means of verifying quickly that the "bread dough is well matured" (the calculated wave function is well reliable). As well as the dispersion and the DOS, the **lattice constants** and **bulk modulus** are also such quantities for the quick check. We didn't explained these quantities because, in this most basic course, we assumed fixed structures to start the calculations. In practice, however, one often performs the structure optimization (as briefly explained later in Sect. 5.1.2), where examining if **the optimized lattice constants and the modulus are consistent with the experimental values** would be a useful check. Without such an explanation, a simple question may arise from a beginner, "Why are we calculating such quantities when we are not interested in lattice constants or bulk moduli?" These quantities are actually the first thing to be checked when a simulation is performed.

The dispersion and DOS are the most likely to be reported in previous studies, so the most appropriate for quickly comparing the obtained results to see if the shapes are consistent or not. If the shape is significantly different, or if the gap does not appear even though it is an insulator, then it is likely to have some mistake in your calculation somewhere. You may be concerned that you are trying to calculate a system for which there is no previous research and no reference to be compared. In this sense, it is a strategy to start not with your final target but with a similar system for which the references by the previous studies are available, as described in Sect. 1.3.2 (also detailed in Sect. 6.1).

While the comparisons among DFT theoretical values do not have any difficult point, we need to be careful when the comparisons are made between theoretical and experimental values. The bandgap estimation is the typical problem, where it is well known that DFT generally gives an underestimation when compared to the experiments (several plausible explanations have also been established as explained in Sect. D). It is also well known for experts that the lattice constants tend to be over-estimated or underestimated depending on the choice of computational conditions. There are often situations where experts say, "it's weird to get such a good coincidence between DFT and experimental values using that computational condition. It **should** give discrepancy...".

Tea Break

However, beginners are never familiar with this kind of knowledge. As such, they tend to waste of a lot of time straggling to get the agreement (never occurs!) by changing the conditions and recalculating blindly. When I checked their situation, I just found there was nothing wrong, and the calculation **should give** the underestimation. This reminds us of the need to have a channel to consult with experts.

References

1. "Solid State Physics", N.W. Ashcroft and N.D. Mermin, Thomson Learning (1976).
2. https://www.quantum-espresso.org (URL confirmed on 2022.11)
3. "Electronic Structure: Basic Theory and Practical Methods", Richard M. Martin, Cambridge University Press (2004/4/8), ISBN-13:978-0521782852
4. "Linux Command Line and Shell Scripting Bible", Richard Blum, Christine Bresnahan, Wiley (2021/1/13) ISBN-13:978-1119700913
5. "Ugokashite Rikaisuru Daiichigenri-denshijoutai-keisan" (in Japanese), Ryo Maezono, Tom Ichibha Morikita Publishing (2020/9/19) ISBN-13:978-4627818217

Part II
Toward Understanding Theoretical Background

This part is designed to help you understand "why" the input file settings given 'by a teacher without any explanations' Chap. 3 are the way they are.

This part is also structured in the way that "practice first, theory later", so that you can learn the later theoretical part with a concrete image of corresponding operations. First, you will experience how the final predictions are affected by the settings in the input file (Chap. 4). After that, the minimum explanations on the background theory is given to understand why such dependence appears (Chap. 5).

Chapter 12 derived from this chapter provides an explanation on the theoretical background for an advanced content. However, the content (DFT+U method) has recently become a matter of common use even for practitioners.

Determining Computational Conditions

4

Abstract

In the previous chapter, we first learned about a series of operations, assuming that the "specifications given to the input file" were "prepared and given by the teacher" in order for you to grasp the "sequence of operations". Now, the main subject of this chapter is to learn how those specifications should be determined in the first place. First, we will explain why it is important to determine the conditions for calculations in the prediction of physical properties. Next, we describe how to determine the parameters those are relatively easy to understand, namely those to define the resolution of the calculation, and give an explanation so that the reader can imagine the physical meaning of the parameters. In the latter part of the lecture, we will understand how much of a difference in prediction is caused by the selection of pseudo-potential and exchange-correlation potential through practical use. The explanation of the reasons for the differences is given in the next chapter (Chap. 5). As an important technical matter, we will explain how to use a script to perform a loop process to automatically submit calculations with different calculation conditions one after another.

4.1 Why the Determination of Calculation Conditions is Important?

4.1.1 To Get Reliable Predictions

Among the specifications that should be included in the input file, the ones that greatly affect the computational prediction are the subjects of this chapter. Even by a beginner, it is easily imagined that the mesh accuracy for integration would matter. It is easy to understand that the finer the mesh gets, the better reliability is achieved. In this case, it is natural to perform several calculations with different mesh coarseness, and determine the reliable value as the converging value as we make the mesh finer.

> **⚠ Attention**
>
> We cannot neglect such a "convergence check" and present a one-shot result, saying, "Maybe it's OK like this". This sense is understandable basically by whoever working in science with whatever one's major is. However, it is indeed the situation for the one who is completely unfamiliar with ab initio calculation, and any meaning of the input parameters is completely not clear to the one. If one gets a finite answer which is in not so strange range after the normal termination of calculations, one would tend to fall into the trap of thinking "well, this is the answer that ab initio calculation gave me". There are actually many beginners who use the same parameter setting as those given in the example file to any practical systems since they think "it is safe not to touch".

In addition to the mesh fineness, there is another fineness about the expanding basis functions. It should carefully be examined to determine at what point one can truncate the expanding series. In order to speed up the convergence, one can adjust how sensitive the iterative update depending on the previous results in the last iteration step. Making it too insensitive, the convergence itself gets better, not so flapped by the previous loop result, but the converged result is suffered from the bias. In such cases, it is still necessary to compare and extrapolate the results with several different choice of the parameter (Sect. 10.2).

In addition to these numerical settings, it is also necessary to consider how much the choice of "the modeling of the interaction between electrons" and "the modeling of the influence of the nucleus felt by the electrons" affects the final prediction. We will explain this in the latter half of this chapter. Unlike the parameters about the resolution treated in the first half of the chapter, there is no simple and clear guideline how to determine the proper choice for these parameters treated in the latter half, such like "the finer, the better". It is a difficult problem that there is no clear answer to the question "then what is the appropriate choice?". Nevertheless, various findings have been accumulated by experts, saying "that selection is obviously not consensus". Anyway, for the parameter choices, authors of a paper should provide some plausible comments on "why we take this choice" on the paper.

In any case, any "publishable level" works require the reliability of the calculated predictions as much as possible. For this purpose, it is necessary to understand and master the procedure to determine appropriate parameter choice as explained in this chapter.

4.1.2 Applying Identical Condition to Predict Series Trend

Predicting a trend of properties when a series of modification is made on a material is one of the major purpose of ab initio analysis with its advantage. In Sect. 1.3.2, we cited the concept of ΔSCF to explain that it is not a good idea to consider ab intio calculations as a tool to predict absolute values of the properties. d Reasonable usage of the tool is to predict trends over a series of comparisons, such as "Which atomic substitutions can surpress thermal expansion" or "What metallic element can be used to make electrodes to achieve anti-peeling strength" [1] etc.. To perform such trend predictions, the **identical conditions** for the computational evaluations should be applied for a "fair comparison", as a matter of course.

In the next and subsequent chapters, we will explain the scheme of how to choose computational parameters for each target. Some beginners might be so impressed by this procedure and considering this as a routine that they misleadingly apply each best choice **independently** to each system in a series to be compared, leading to "an unfair comparison" with not identical conditions. As we explain from the next chapter, "the best choice" means such a condition to achieve cheaper costs as much as possible with keeping required accuracy (deviation from the converging value). When forecasting a trend, it is important to employ the most stringent conditions among several systems to maintain the quality of bias at the same level.

Tea Break

When I collaborated with external group with a beginner student, he found an interesting trend to predict a best choice of the element to be doped to improve a property. I examined his calculation log (the input/output file left behind) for secure, and found that the computational conditions were different for each systems (best choices for each systems). Recalculating the trend, it turned out to give completely different trend predictions.

4.2   Parameters for Accuracy via Resolution

For the current tutorial using DFT applied to periodic systems with plain wave basis set expansion, the representative parameters to control the accuracy via describing resolution are the k-mesh and E_{cut} (cutoff energy). These are the essential specification that should be clarified on a paper to ensure the reproduction of results by others.[1]

4.2.1 k-Mesh

As the word "mesh" implies, this parameter defines the coarseness of the mesh when discretizing the integral. When we mention a $3\times3\times3$-mesh, it means the numerical

---

[1] The readers might include those mainly using Gaussian basis set for isolated systems like molecules. For isolated systems, there is no such concept like k-mesh, basically, but we can regard it as a calculation with $1\times1\times1$-mesh (sometimes referred as "Γ-point calculation"). What is corresponding to E_{cut} is the basis set quality specification such as "6-31G", "6-311G++", etc. [2], for which we will explain more in detail in the next chapter.

evaluation by taking $3\times3\times3$-grid for the region to be integrated in k-space. As you can immediately understand intuitively, the finer, the more descriptive it is. Figure 4.2 shows the energy getting converged as K increased for a mesh of $K \times K \times K$ with keeping other conditions to be the same. For a large enough K, the results will converge without much change beyond that. It is also easy to imagine that "the finer the mesh, the higher the computational cost. As such, the optimal K is determined as coarse the mesh as possible so that the result reaches the converged region.

The above is the way to explain it from a numerical viewpoint. On the other hand, it is important to consider the physical meaning of the k-mesh so that you can have a concrete image of it when adjusting it. The "k" in k-mesh refers to the wavenumber $\mathbf{k}$ appearing in the basis function expansion of the wave function in a plane wave as

$$\psi(\mathbf{r}) \sim \sum_{\mathbf{k}}^{|\mathbf{k}_{max}|} \psi_{\mathbf{k}} \cdot \exp[i\mathbf{k} \cdot \mathbf{r}]. \tag{4.1}$$

Taking a width of the mesh Δk, the corresponding length λ_{max} given by the relation, $\Delta k \sim 1/\lambda_{max}$, is the maximum wavelength that can be described by the mesh resolution. "Using a finer mesh" therefore means "incorporating even longer wavelength components of the wave function". Thus, the convergence in Fig. 4.2 means that "longer wavelength components beyond a certain level are not relevant even when they are taken into account". In other words, "taking a coarse k-mesh" means "how much of the longer wavelength component can be truncated".

As explained at Fig. 11.3 in the appendix chapter, the two-partition mesh of the Brillouin Zone ($\Delta k = 2\pi/2$) can describe twice longer wavelength of the lattice parameter, the three-partition mesh of the Brillouin Zone ($\Delta k = 2\pi/3$) can describe three times of the lattice parameter, ... This means that a $2 \times 2 \times 2$-k mesh calculation is taking a simulation cell that is a multiple of $2 \times 2 \times 2$ of the unit cell of the considered structure.

4.2.2 Cutoff Energy

For the cutoff upper limit appeared in the summation of Eq. (4.1), $|\mathbf{k}_{max}|$, we can consider the squared value $|\mathbf{k}_{max}|^2 \sim E_{cut}$ with the energy dimension.[2] This is called **cutoff energy** in the expansion.

The larger E_{cut} means the larger number of the expanding terms, leading to the **finer description of the behavior** of the wave function. It also leads to higher computational cost as a tradeoff. Figure 4.1 shows the dependence of the energy on the choice of E_{cut}. Again, as in the case of the k-mesh, the result converges at a certain level of E_{cut}, and any larger value will only increase the computational cost. Therefore, as in k-mesh, the optimal E_{cut} is determined as smallest possible so that the energy reaches the convergence region.

---

[2] It is the convention in ab intio calculations to take $m_e = 1$ and $\hbar = 1$, called natural unit system.

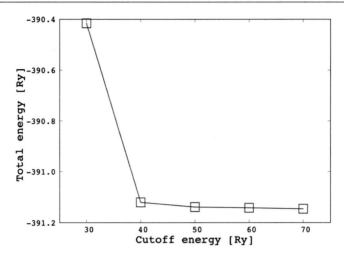

Fig. 4.1 SCF energies (in Ry unit, given as the vertical axis) are converging as the cutoff energy E_{cut} (in Ry, as the horizontal axis). The results are evaluated for the α-Quartz structure. The optimal value for E_{cut} is chosen as least as possible in the converging range (50 Ry in this example)

For E_{cut} as well, it is important to have an image of its **physical meaning**. E_{cut} is a parameter that truncates the expanding component above $|\mathbf{k}_{max}|$. Since wavenumber and wavelength are inversely related, "truncating larger wavenumber" means "truncating shorter wavelengths". Therefore, taking smaller E_{cut} to save the computational cost corresponds to the truncation of the shorter wavelength component based on the observation that such a shorter length does not matter to describe the target phenomena.

4.2.3 Physical Image of Truncation

The E_{cut} and k-mesh are explained as corresponding to shorter and longer wavelength truncations, respectively. By keeping these meanings in mind, you will be able to get a sense and justification of what you are approximating by the truncation in view of the characteristics of the system you are working on. First, for **shorter wavelength side** truncation (E_{cut}), the situation requiring the shorter wavelength components occurs when the considered wave function oscillates drastically. The wave function varies smoothly in the valence region, but oscillates drastically in the vicinity of the nucleus [4]. How drastic the oscillations is determined by the pseudopotential to be adopted (Sect. 6.3.3). Wavelengths shorter than the characteristic length of the oscillation no longer affect the descriptiveness, and then truncated by the proper choice of E_{cut} so as not to increase the computational cost.

Next, for the **longer wavelength side** truncation (k-mesh), it matters to how coherent the considered wave function is, namely, how far the phase of the wave function can be transmitted without being chopped up. As explained in the appendix (Sect. 11.5.3), the distinction between the metallic and the insulating properties can be understood in terms of the coherence [3]. When the coherence is propagating throughout the crystal, it is a metal, while in insulators it is chopped up. The longer wavelength components are therefore required for describing metallic systems, leading to the denser k-mesh. This consequence based on the consideration by **real-space image** can also be concluded from another viewpoint, namely, from the **reciprocal space** image. In the reciprocal space, the electrons are filled from the lower end of the band within the Brillouin zone (Sect. 11.4). The structure in k space assembled in this way is the integration region for calculating the electronic state [4].[3] In the case of an insulating state, where the upmost band is completely filled up and the next band is empty, the shape of the filled up structure matches the shape of the Brillouin zone, so the polyhedron retains a certain degree of simplicity. On the other hand, for a metallic state, the band is filled partially upto the Fermi level forming a complex shape of the "water surface" (=Fermi surface). "The structure in k-space", (i.e., the region to be integrated over) has therefore a complex shape as well, as if the Brillouin zone has been eroded to a complex Fermi surface [4]. To integrate such a complex shape accurately with an integral mesh, a coarse mesh will miss finer characters such as small protrusions. This is a usual explanation why the finer k-mesh is required for metallic systems based on **the reciprocal space picture**. It corresponds to the above discussion based on the real space picture via the reciprocal relationship of $\lambda \sim 1/k$.

In the case of an isolated molecular system using a Gaussian basis set [2], the choice of a basis set corresponds to the adjustment of the E_{cut} and k-meshes here. Adopting a better basis set corresponds to the choice that can describe more finer behavior around the origin of the orbital function (corresponding to the shorter wavelength side), or that can describe broader tails of the orbitals (corresponding to the longer wavelength side). For Gaussian basis sets, it is relatively difficult to become proficient level being capable to take optimal choice of a basis set for a given system[4] because of a lot of degrees of freedom to tune the basis set. In this aspect, the plane-wave basis has simpler degrees of freedom, tuned just by a parameter, E_{cut}, which is very helpful for beginners.

4.3   Procedure to Determine the Resolution Specifications

To determine the conditions for computational resolution, k-mesh size and E_{cut}, it would be recommended to start with E_{cut} first, and then decide on the k-mesh size. If we try to do it the other way around, we are faced with the question, "How much E_{cut} should we take for the time being when looking at the convergence of k mesh size?". As mentioned above, E_{cut} is defined by observing how drastically the wave

---

[3] When calculating physical quantities, it is necessary to count up all the electronic states occupying the band, which is performed as the integration over the region of the structure in k-space assembled from the occupied states.

[4] In order to be able to handle the practical level, one needs to be familiar with several correction schemes related to the overlapping integral error between bases (BSSE), the complete basis limit (CBS) etc.

function oscillates near the nucleus. This is greatly depending on the characteristics of the target system and the pseudopotential used, and it is not easy for a beginner to answer in terms of the scale in E_{cut}.

On the other hand, the k-mesh is rather easier to imagine as it is the measure of upto how long the wavelength is required to be described for the wave function in the unit of the lattice constant of the unit cell [e.g., "$2 \times 2 \times 2$" is the condition to take into account upto twice longer wavelength, and larger wavelength are truncated to be described]. Therefore, it would be reasonable to take $2 \times 2 \times 2$ for the time being, and then examined the convergence of E_{cut} to determine it. Once E_{cut} has been fixed, then we can check the convergence of k-mesh by changing it from around $2 \times 2 \times 2$.

4.3.1   Determining Cutoff Energy

During the explanation about the procedure to determine E_{cut}, we also explain about the **script processing** to handle the input file in a programming manner, that is frequently used in subsequent processes.

Let us prepare the working directory, "02calibCond", for the present tutorial as

```
% pwd
    /home/student/work
% mkdir 02calibCond
% cd 02calibCond/
% cp ~/work/01tryout/scf.in ~/work/02calibCond/ecut.in
% cp ~/work/01tryout/*.UPF .
```

where we copied the input file and pseudo potentials into there.

Inside of the input file, you can see the parameter "ecutwfc",[5] that is specifying the value for E_{cut}. We would like to change the value of "ecutwfc" as 30, 40, ... upto 70 with the interval by 10,[6] and to calculate the energy depending on "ecutwfc" as in Fig. 4.1 to see the convergence.

It is quite tedious to perform this in the form of "manually changing input parameters and performing SCF calculations individually". Therefore, we will learn how to use **shell script for automatic handling** of input files, as described below. Scripting work as introduce in Sect. 3.4.2 is also possible to perform **loop processing** as explained below.

Let us begin with editing the input file "ecut.in"[7] used for a SCF calculation as

---

[5] This is the mnemonic based of "E_{cut} of underline{wave}underline{fun}underline{c}tion'.

[6] The unit for ecutwfc is "Ry" (Rydberg).

[7] To avoid typos, the input files used in this course are included as text files in the downloaded materials as introduced in Sect. 2.4.2. You can copy them, or "cut and paste", or for comparing to find mistakes using "diff" or "sdiff" command. These files are stored with the same file name under "~/work/setupMaezono/inputFiles/".

```
. . .
&SYSTEM
. . .
ecutwfc =    target1
ecutrho =    target2
/
. . .
```

Instead of specifying a concrete value for "ecutwfc", we put a string "target1" for it. The string will be replaced by a concrete value later handled by a shell script, repeatedly [5].

The same strategy is applied also to another parameter "ecutrho"[8] with the string "target2". The "ecutrho" is the cutoff energy for the plain wave expansion not for the wave function but for the charge density. For PAW pseudo potentials (what we are using), it is recommended to take "ecutrho" as 12 times of the value of "ecutwfc" [6], for which we follow the recommendation.[9]

The script to replace the string "target1" and "target2" is prepared as[10]

```
% cat ecut.sh
    #!/bin/bash
    for ecut in 30 40 50 60 70;
    do
        target1=$ecut
        target2=`expr 12 \* $ecut`
        title=ecut_$ecut
        cat ecut.in | sed -e "s/target1/$target1/g" \
            -e "s/target2/$target2/g" > $title.in
        pw.x < $title.in | tee $title.out
    done
```

In this script [5], we note the followings,

- **Declarative statement**
  "#!/bin/bash" appearing at the beginning of the script corresponds to the declaration that "This file is a shell script".

---

[8] This is the mnemonic of "'ecut' of rho (= ρ; charge density).

[9] This is a typical example of an input parameters that are "too advanced for beginners for the time being and can be left for later understanding" as described in Sect. 5.2.1. It is best to use the recommended values, leaving it to the methodological experts to decide what settings are best. However, the recommended values are often updated as time goes by, so you need to pay attention to such updating.

[10] The backslash "\" is used to show a too long string to fit on a line (see Acronym).

- **Processing by sed**
  One can make "sed" perform more than one process by using the option "-e" as '-e "process01" -e "process02"'.
- **for-loop structure**
  The processes written between 'do' and 'done' are performed repeatedly as loop wise. The head line with "for" specifies the variable "ecut" to be substituted by the values 30, 40, ..., 70, sequentially in the loop execution. The value assigned to "ecut" is cited as the variable "$ecut".
- **Arithmetic operation using "expr"**
  At the line with "target2", the arithmetic operation is performed using **expr** command. You can easily check on the command line that "% expr 12 * 3" returns "36" as "*" is used for multiplication instead of the usual "*". Note that there should be a space on both sides of an arithmetic operation symbol.
- **Substituting the command output**
  The format, "target2 = '(command)'", substitutes the output value of the command execution into the variable 'target2'. Note that the **backquote** is used to enclose the (command), not the single quote.

Keeping the above notes in mind, you can understand how the 'ecut.sh' works as

- The script generates a series of input files, 'ecut_30.in', 'ecut_40.in', ... based on the seed file 'ecut.in', replacing the string 'target1' by the value of 'ecutwfc' as 30, 40, ...
- In these input files, 'ecut_30.in', 'ecut_40.in', ..., the value for 'ecutrho' is set as the 12 times of the value of 'ecutwfc'.
- The script executes a series of SCF calculations using 'ecut_30.in', 'ecut_40.in', ..., by its loop structure, writing out each output in the output files 'ecut_30.out', 'ecut_40.out', ...

Then, execute 'ecut.sh' as

```
% bash ecut.sh
```

It will take around 30 min. After all the calculations are completed, you can check the convergence of the energies using 'grep' as

```
% grep ! ecut_*.out
    ecut_30.out:!    total energy    =    -390.41574965 Ry
    ecut_40.out:!    total energy    =    -391.12042402 Ry
    ecut_50.out:!    total energy    =    -391.13887557 Ry
    ecut_60.out:!    total energy    =    -391.14192878 Ry
    ecut_70.out:!    total energy    =    -391.14594216 Ry
```

To extract the data for plotting the energy dependence on the E_{cut}, execute a command composed of 'sed' and 'awk' as

```
% grep ! ecut_*.out | awk '{print $1, $5}' \
    | sed -e 's/ecut_//g' -e 's/.out:!//g'
    30 -390.41574965
    40 -391.12042402
    50 -391.13887557
    60 -391.14192878
    70 -391.14594216
```

By making a redirect, '> ecut.dat', we can save the above data in the text file 'ecut.dat'.

By using 'gnuplot' as

```
% gnuplot
    gnuplot> plot 'ecut.dat' u 1:2 w lp
```

we can get the plot as shown in Fig. 4.1.

! Attention

The following operation needs to be done directly on the RaspberryPi because of the handling of graphical pictures. If you are operating it remotely from your familiar PC (Sect. 2.3.3), please switch to login directly to the RaspberryPi now (it is reminded later at the point where you can get back to the remote connection).

In Fig. 4.1, we observe the SCF energy convergence as E_{cut} increases at around 40–50 Ry. We will take $E_{cut} = 50$ Ry as the optimal choice.

For the criterion of the convergence, it is a common choice to take 'the deviation within 0.001 [hartree] (= 0.002 [Ry]) from the converged energy value. The deviation 0.001 [hartree] $\sim$ 1 [kcal/mol] is known as **the chemical accuracy** [2], which is the typical resolution required to distinguish between a single bond and a double bond. Looking at the grep output above, we see that the chemical accuracy is not achieved even at 70 Ry, but we chose 50 Ry as the optimal value used for E_{cut} here after. A justification for this choice is based on the fact that what matters is not the absolute energy value for each phase but the energy difference between phases. The choice of 50 Ry was actually confirmed to achieve the chemical accuracy for the energy differences between any polymorphs in this case.

4.3.2 Determining Optimal k-Mesh

Fixing E_{cut} to 50 Ry, we next examine the convergence with respect to the k-mesh to determine the appropriate mesh size. The input file 'ecut_50.in' generated in the previous subsection is copied and used under the name 'kmesh.in'.

In 'kmesh.in', the block defining the k-mesh size is found at

```
K_POINTS {automatic}
2 2 2 1 1 1
```

where '2 2 2' specifies $2 \times 2 \times 2$-mesh. Then, we would like to change this from '1 1 1' to '4 4 4' in increments of one, and examine the convergence of the total energy with respect to the mesh size. As it is tremendous to do this manually, we apply the script work as explained in the previous subsection.

As we did in the previous subsection, we replace the target part to be modified, '2 2 2' in this case, by a string 'target' as

```
...
K_POINTS {automatic}
target target target 1 1 1
...
```

Then prepare the script 'kmesh.sh'[11] as

```
#!/bin/bash
for kt in 1 2 3 4;
do
    title=kmesh_$kt
    cat kmesh.in | sed -e "s/target/$kt/g" > $title.in
    pw.x < ${title}.in | tee ${title}.out
done
```

You are now expected to understand what the script is doing, if you understand the contents in the previous subsection. Then, execute the script as

```
% bash kmesh.sh
```

It generates a series of input files, 'kmesh_1.in', 'kmesh_2.in', $\cdots$, for mesh sizes, $1 \times 1 \times 1$, $2 \times 2 \times 2$, $\cdots$, executing each SCF calculation to be output as 'kmesh_1.out' etc. (it will take around an hour).

! Attention

The following operation needs to be done directly on the RaspberryPi because of the handling of graphical pictures. If you are operating it remotely from your familiar PC (Sect. 2.3.3), please switch to login directly to the RaspberryPi now (it is reminded later at the point where you can get back to the remote connection).

---

[11] To avoid typos, the input files used in this course are included as text files in the downloaded materials as introduced in Sect. 2.4.2. You can copy them, or 'cut and paste', or for comparing to find mistakes using 'diff' or 'sdiff' command. These files are stored with the same file name under '~/work/setupMaezono/inputFiles/'.

Fig. 4.2 The dependence of
the total energy (vertical, in
unit of Ry) on the *k*-mesh
size (horizontal). The results
are evaluated for the α-quartz
case. As the finer mesh
requires more expensive
computational cost, the
optimal choice is taken as
least as possible among those
satisfying the convergence
($2 \times 2 \times 2$ in this case)

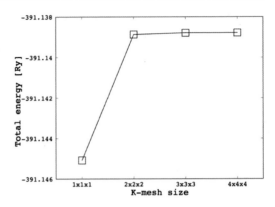

To look at the dependence of the total energies on the *k*-mesh size, we can extract
the data by[12]

```
% grep ! kmesh_*.out | awk '{print $1, $5}' \
  | sed -e 's/kmesh_//g' -e 's/.out:!//g'
    1 -391.14507980
    2 -391.13887557
    3 -391.13879391
    4 -391.13879107
```

Putting the redirect to 'kmesh.dat', we can plot the dependence by using 'gnuplot'
as

```
% gnuplot
    gnuplot> plot 'kmesh.dat' u 1:2 w lp
```

as shown in Fig. 4.2. In the figure, we can see tha the total energy gets converged at
around $2 \times 2 \times 2$ mesh size.

4.4   Choice of Pseudopotentials

4.4.1   Specification of Pseudopotentials

We have briefly introduced the concept of pseudopotential in Sect. 3.2.4. Pseudopotentials effectively describe the potentials from nuclei felt by valence electrons represented as a numerical potentials. There have been many different schemes developed

---

[12] The backslash '\' is used to show a too long string to fit on a line (see Acronym).

how to create the numerical potentials. 'Fabrication of pseudopotentials' forms a substantially specialized research field. Even experts in electronic structure calculations usually consider the pseudopotentials as 'a black box' which is prepared by the experts, use it just by downloading from a library on the Web.

Nevertheless, there is a minimum level of knowledge about the specs of the pseudopotentials that should be acquired even for the user level. As we will see in the next chapter, it is important to recognize well

- how the core/valence separation is made in your calculations?
- which type of pseudopotentials you are using (norm-conserved one, or ultrasoft, or PAW)?

For these minimum information, even the user should be prepared to answer such questions at conference talks, and these information should clearly be stated in your paper when your calculations are reported. It might be better to add the required knowledge for a user about if your pseudopotentials are taking into account the relativistic effects or not (and to what extent). It's perfect for a user level if one could recognize that the pseudopotentials are generated and used for **which valence state** to be expected. For more detailed specs such as 'Troullier-Martins type', 'KB separation' etc. can be regarded beyond the user-level knowledge, leaving it to specialists.

From the next subsection, we will experience how much the prediction depends on the choice of pseudopotentials. For this tutorials, we compare two different type of pseudopotentials, ultrasoft-type and PAW-type with the same valence/core separation.

4.4.2 Dependence on the Pseudopotential Choice

Let us remember that we have prepared two different types of pseudopotentials in Sect. 3.2.4. For silicon atom, these were 'Si.pbe-n-kjpaw_psl.1.0.0.UPF' and 'Si.pbe-n-rrkjus_psl.1.0.0.UPF'. We can find the string 'paw' and 'us' before '_ps', each of which means PAW (Projector Augmented Wave) and US (Ultra Soft). You can also find the same for oxygen pseudopotentials you have downloaded.

By using these two types of pseudopotentials, we examine the dependence. Let us prepare the working directory as

```
% pwd
    /home/student/work
% mkdir 03pseudo
% cd 03pseudo/
% cp ../structure/quartz_alpha.in qa_scf.in
% cp ../structure/quartz_beta.in qb_scf.in
```

where we prepared the input file for α (β)-quartz as qa_scf.in (qb_scf.in).

By using 'emacs', edit your input files so that the k-mesh to be set as $2\times2\times2$, and E_{cut} (ecutwfc) to be 50 Ry. For 'ecutrho', it is set as 12 times of 'ecutwfc' being 600 Ry as we explained earlier:

```
...
ecutwfc=50
ecutrho=600
...
K_POINTS {automatic}
2 2 2 1 1 1
...
```

Then, copy all four pseudopotentials stored under 'work/pp' to the working directory:

```
% pwd
    /home/student/work/03pseudo
% cp ../pp/* .
% ls
    qa_scf.in   qb_scf.in
    O.pbe-n-kjpaw_psl.1.0.0.UPF
    O.pbe-n-rrkjus_psl.1.0.0.UPF
    Si.pbe-n-kjpaw_psl.1.0.0.UPF
    Si.pbe-n-rrkjus_psl.1.0.0.UPF
```

Using the 'qa(b)_scf.in' as the prototype, prepare each input files for each pseudopotential choice as

```
% cp qa_scf.in qa_paw_scf.in
% cp qa_scf.in qa_us_scf.in
% cp qb_scf.in qb_paw_scf.in
% cp qb_scf.in qb_us_scf.in
```

Since the prototypes are originally for PAW-type pseudopotentials, we have to amend only 'qa(b)_us_scf.in' so that their specifications for pseudopotentials can be set for US-type as

```
...
ATOMIC_SPECIES
   Si    28.08500    Si.pbe-n-rrkjus_psl.1.0.0.UPF
    O    15.99900     O.pbe-n-rrkjus_psl.1.0.0.UPF
ATOMIC_POSITIONS {crystal}
...
```

Having prepared as above, then we can execute each calculation as

```
% pw.x < qa_paw_scf.in   | tee qa_paw_scf.out
% pw.x < qa_us_scf.in    | tee qa_us_scf.out
% pw.x < qb_paw_scf.in   | tee qb_paw_scf.out
% pw.x < qb_us_scf.in    | tee qb_us_scf.out
```

Having completed all the calculations, extract the final converged energies from each output file. Since each final value is written at the line starting with '!', we can extract the values using the wildcard '*' as

```
% grep ! *.out
    qa_paw_scf.out:!  total energy = -391.13887583 Ry
    qa_us_scf.out: !  total energy = -229.67899532 Ry
    qb_paw_scf.out:!  total energy = -391.12397608 Ry
    qb_us_scf.out: !  total energy = -229.66412806 Ry
```

We can see that the values by PAW and US are quite different. To analyse this, we first check whether they adopt the same valence/core separation,

```
% grep 'number of electrons' *out
    qa_paw_scf.out:    number of electrons   =   48.00
    qa_us_scf.out :    number of electrons   =   48.00
    qb_paw_scf.out:    number of electrons   =   48.00
    qb_us_scf.out :    number of electrons   =   48.00
```

As we confirmed, there is no difference on the number of valence electrons (=48) in this example case.[13]

It might surprize the beginners that the energy values are so different between the choice even with the same number of valence electrons. This can immediately be understood due to the difference in 'the origin of energy zero' for each pseudopotentials. We can remember that 'the reference zero' for the energy does not matter but what matters for the physics is 'the difference' between the energies.

So what about the difference? The difference which matters is that between the α- and β-quartz energies as predicted. Taking the subtraction, the differences have been evaluated as

```
PAW: 0.01489975 Ry
 US: 0.01486726 Ry
```

achieving a quite good in agreement each other.

---

[13] In general, it can be different, so we have confirmed this point first.

By the way, there is no need to launch your 'calculator' application or 'Excel' to calculate the above differences. The differences are evaluated from

```
PAW     -391.13887583 +391.12397608
US      -229.67899532 +229.66412806
```

Putting the above values on your editor using 'cut and paste', editing with plus and minus signs, then put the string on your terminal after the 'bc' command (Sect. 9.2.2) is executed using 'cut and paste' again,[14]

```
% bc
    -391.13887583 +391.12397608
    -.01489975
    -229.67899532 +229.66412806
    -.01486726
```

To quit from 'bc', type 'Ctrl+D'.

> **Important**

It is not a good practice for the research log, just recording 'US:-0.01486726 Ry'. It is much better to record it as '$-229.67899532 + 229.66412806 = -0.01486726$' by 'cut and paste' on the log. By doing that, it is easier to trace back how to get this result to find the mistake (possibly occurring during the unit conversions).

The present example is a quite beautiful case typically showing that the relative differences are well coinciding each other even each absolute values are quite different. For the same number of valence electrons, it should, in principle, be expected to get almost the same result even using different pseudopotentials. However, there are such problems where no unique consensus is possible how to devide core and valence electrons (Sect. 5.5.1). When one examine the energy difference between two phases or polymorphs using different pseudopotentials with different core/valence separation, one will get a discrepancy about the predictions in general. This is because the contribution to the binding from the valence electrons can differ each other if one uses different core/valence separations. In such problems exhibiting predictions sensitive

---

[14] We strongly recommend to use any of 'text editor' instead of 'MS-Word'. Most of beginners (including the author in past!) tend to write everything in Word files (as recognized as 'files'), but since Word is originally a multifunctional **word processing software** for creating beautiful 'final documents' including figures, variations of fonts etc., it is heavy in operation and capacity. If you are handling only the text information, no need to use such a heavy operation tools at all. You will find how lighter and handy operations are realized by the text editors. You can find any of your appropriate text editors working on your OS (usually free, long-life, and world-wide users) by search on web as 'text editor windows' etc. Note for the beginners that it is possible to save and keep files even for text editors, as a matter of course. It is never '(Document files) = (Word files)'.

to the core/valence separation, the contributions from 'semi-core' electrons (they can be valence or core depending on the choice of the separation) are not negligible [8].

4.5 Choice of Exchange-Correlation Potentials

4.5.1 Starting from Rough Description

The exchange-correlation potentials[15] are the most essential part in DFT, as explained in detail in the next chapter (Sect. 5.3.3). A user has to determine which exchange-correlation potential to use by specifying in the input file. However, if a beginner tries to understand what the exchange-correlation potential is, one will find a lot of theory behind it. It is not easy to find a direct answer to the question 'Which exchange-correlation potential should I choose'.

To put it bluntly, the choice of the exchange-correlation potential corresponds to the choice of the model[16] of quantum inter-particle interaction effects. It is not possible to give a definitive answer for a recommended choice, because it is still at the level of fundamental research to understand what kind of exchange-correlation potential is suitable for the system in question. A safer choice would be to adopt the same exchange-correlation potential that has been used in preceding works for 'similar' systems as your target. Here, 'similar' means the similar nature of the bonding (ionic, covalent, intermolecular etc.) in the system in question.

4.5.2 Dependence on Exchange-Correlation Potentials

Putting the theoretical introduction of it afterward in Sect. 5.3.3, let us experience how the predictions are affected by the choice of exchange-correlation potentials. We adopt several different choice from those typically used ones [7], PZ(LDA), PBE, PW91, and BLYP, to evaluate the energy differences between four polymorphs of silicon oxides (α (β)-quartz and -cristobalite).

Let us prepare the working directory as

```
% pwd
    /home/student/work
% mkdir 04stability
% cd 04stability/
% cp ../03pseudo/*paw*.in .
% mv qa_paw_scf.in qa_scf.in
% mv qb_paw_scf.in qb_scf.in
```

---

[15] It is also called exchange-correlation functionals depending on contexts.

[16] In this wording, we note again that the meaning of the 'model' here is a slightly different from that used for the model theories. It will be explained in detail in the next chapter (Sect. 5.3.3).

```
% cp ../03pseudo/*paw*.UPF .
% cp ../structure/cristo_alpha.in ca_scf.in
% cp ../structure/cristo_beta.in cb_scf.in
% ls
    ca_scf.in   cb_scf.in
    qa_scf.in   qb_scf.in
    O.pbe-n-kjpaw_ps1.1.0.0.UPF
    Si.pbe-n-kjpaw_ps1.1.0.0.UPF
```

where we only used PAW pseudopotentials. For the names such as 'qb_scf.in', 'a' and 'b' represent α and β, while 'c' and 'q' mean cristobalite and quartz, respectively.

Then, amend the contents of each input file ('qb_scf.in' etc.) at the lines with the comment, '! attention(1) or (2)':

```
&control
calculation='scf'
restart_mode='from_scratch'
prefix='qa'
outdir='out_qa'
pseudo_dir='./'
/
&SYSTEM
   ibrav = 0
   !A =      4.91400
   nat = 9
   ntyp = 2
   occupations='fixed'    ! attention(1)
   ecutwfc=50             ! attention(1)
   ecutrho=600            ! attention(1)
   input_dft='XC_FUNC'    ! attention(2)
/
&electrons
/
K_POINTS {automatic}
2 2 2 1 1 1              ! attention
...
```

At the lines with '! attention(1)', we amend the specifications about the resolutions so that the same conditions are adopted for different choice of exchange-correlation potentials for comparison. At the lines with '! attention(2)', the parameter 'input_dft' specifies the choice of exchange-correlation potentials. We are putting the string 'XC_FUNC' as the variable which will be replaced as 'PBE', 'LDA', 'BLYP', and 'PW91, by the script as we learned in the preceding sections.

The script is prepared as follows:

```
% cat run.sh
  #!/bin/bash
  for xc in pz pbe blyp pw91; # functionals
  do
    for base in qa qb ca cb; # structures
    do
      cat $base'_scf.in' | sed "s/XC_FUNC/$xc/g" \
          > $base'_'$xc'_scf.in'
      pw.x < $base'_'$xc'_scf.in' | \
          tee $base'_'$xc'_scf.out'
    done
  done
```

If you have understood the explanations about the script work given in Sect. 4.3.1, you should be able to decipher what the above script with loop structure is doing. Let us execute it as

```
% bash run.sh
```

which will take around 120 min.

Getting all calculations completed, let us extract the results using 'grep' for quick check:

```
% grep ! *pz*.out
    ca_pz_scf.out:!    total energy = -506.37328143 Ry
    cb_pz_scf.out:!    total energy = -253.13129683 Ry
    qa_pz_scf.out:!    total energy = -379.78693798 Ry
    qb_pz_scf.out:!    total energy = -379.76230793 Ry
% grep ! *pbe*.out
    ca_pbe_scf.out:!   total energy = -521.52258817 Ry
    cb_pbe_scf.out:!   total energy = -260.68669663 Ry
    qa_pbe_scf.out:!   total energy = -391.13887583 Ry
    qb_pbe_scf.out:!   total energy = -391.12397608 Ry
% grep ! *blyp*.out
    ca_blyp_scf.out:!  total energy = -523.79701529 Ry
    cb_blyp_scf.out:!  total energy = -261.82224401 Ry
    qa_blyp_scf.out:!  total energy = -392.83983162 Ry
    qb_blyp_scf.out:!  total energy = -392.82694647 Ry
% grep ! *pw91*.out
    ca_pw91_scf.out:!  total energy = -523.48381613 Ry
    cb_pw91_scf.out:!  total energy = -261.67073035 Ry
    qa_pw91_scf.out:!  total energy = -392.60899012 Ry
    qb_pw91_scf.out:!  total energy = -392.59332372 Ry
```

We notice that values are never comparable among the polymorphs, around twice difference from smallest to biggest. This is because the energies shown above are given in 'per unitcell'. Each unitcell contains different numbers of Si_2 unit for each polymorph, and hence the total values are quite different each other among the polymorphs.

To make meaningful comparison, the values shown above should be divided by the number of Si_2 units contained in each unitcell, giving the energy 'per Si_2 unit' (it is called 'per formula unit'). A way to identify how many formula unit is included in each unitcell is to count up the line with 'Si' in the block of 'ATOMIC_POSITIONS' in each 'scf.in'. It reads 4, 2, 3, and 3 for 'ca', 'cb', 'qa', and 'qb', respectively.

To perform the normalization into 'per formula unit', let us write out the results first as

```
% grep ! *pz*.out > pz.result
```

Then edit 'pz.result' by putting the number (4, 2, 3, and 3) at the end of each line as,

```
% cat pz.result
   ca_pz_scf.out:!    total energy =   -506.37328143 Ry   4
   cb_pz_scf.out:!    total energy =   -253.13129683 Ry   2
   qa_pz_scf.out:!    total energy =   -379.78693798 Ry   3
   qb_pz_scf.out:!    total energy =   -379.76230793 Ry   3
```

Then, the following command gives the normalized values,

```
% awk '{OFMT="%.6f"} {print $1, $5/$7, "Ry/f.u."}' pz.result \
   | sed 's/_pz_scf.out:!//g'
      ca -126.593320 Ry/f.u.
      cb -126.565648 Ry/f.u.
      qa -126.595646 Ry/f.u.
      qb -126.587436 Ry/f.u.
```

where 'f.u' abbreviates 'formula unit' The block 'OFMT="%.6f"' in 'awk' command specifies the number of significant digits to be displayed (up to six decimal places in floating-point style in this case) [5].

Repeat the same normalization to other exchange-correlation potentials (not only 'pz' but also 'pbe', 'blyp', and 'pw91') to get the results,

```
<pz>
     ca -126.593320 Ry/f.u.  2
     cb -126.565648 Ry/f.u.  4
     qa -126.595646 Ry/f.u.  1
     qb -126.587436 Ry/f.u.  3
<pbe>
     ca -130.380647 Ry/f.u.  1
```

```
    cb -130.343348 Ry/f.u.  4
    qa -130.379625 Ry/f.u.  2
    qb -130.374659 Ry/f.u.  3
<blyp>
    ca -130.949254 Ry/f.u.  1
    cb -130.911122 Ry/f.u.  4
    qa -130.946611 Ry/f.u.  2
    qb -130.942315 Ry/f.u.  3
<pw91>
    ca -130.870954 Ry/f.u.  1
    cb -130.835365 Ry/f.u.  4
    qa -130.869663 Ry/f.u.  2
    qb -130.864441 Ry/f.u.  3
```

where the numbers at the end of each line means the order ranked from lowest to highest energy. The results read,

- Three exchange-correlation potentials, 'PBE', 'BLYP', and 'PW91', predict the same order of stabilities for each polymorphs,
- Only 'LDA' predicts the α-quartz as the most stable phase being consistent with experimental fact.[17]

4.6 How the Computational Conditions Affect

4.6.1 How to Read the Results

In the previous section, we saw that only LDA could reproduce the experimental fact (α-quartz being the most stable). Based on this, one might conclude that LDA is the best to describe the system. However, it is too early to conclude that LDA is the most proper choice for this system. We have to be careful to reserve such a superficial conclusion. GGA (PBE) has actually been developed as the *improved* version up over LDA [7]. There are such systems known that both LDA and GGA are not properly working. As a matter of course, there are several systems as well to be reported that LDA gives better predictions. Nevertheless, it is *unknown* issue in general that which exchange-correlation potential is the best to describe the target system properly.

One might refute that LDA actually reproduce the experimental fact, but one have to remember that we performed the comparison for the system with fixed geometry, i.e., lattice constants. I have actually omitted to examine whether the given geometry is the best (i.e., being at equilibrium) for each exchange-correlation potential in this tutorial example. In general, the ground state energy evaluations should be performed

---

[17] From experimental facts, the correct answer is known to be α-quartz ('qa') [9].

for each equilibrium geometry so that it can ensure several consistencies that are satisfied at equilibrium [4, 10].[18]

To get the equilibrium structure for each given exchange-correlation potential, we perform the **geometrical optimization** calculations that is briefly explained in Sect. 5.1.2.

In practical projects, we usually perform the geometrical optimizations for all 16 combinations, [ca/cb/qa/qb] $\otimes$ [pz/pbe/pw91/blyp], to get the equilibrium structure for each choice to be used for the ground state energy evaluations. The results of the energy differences between the polymorphs using each equilibrium geometry will differ from those as we got in the previous section. The comparison between the updated predictions using the optimized geometries would be more plausible to discuss which exchange-correlation potential gives the best coincidence with experiments.

As described in Sect. 5.1.2, the structural optimization calculation itself can be understood as a series of reputations of single-point calculations. To perform these optimizations, it is hence necessary to have adequate computing resources (not only a single fast processing power but also enough number of computing units so that many independent jobs (such as 16 combinations) can be processed simultaneously in parallel wise.

For handling 16 different settings (job submissions, job collections, and summarizing the results to get energy differences etc.), it is quite tedious and likely to be suffered from mistakes if done manually. To avoid this, it is very important (or to say 'essential') to be familiar with the script work to handle them automatically without mistake.

Tea Break

In collaborations the author experienced, the students in the partner side sometimes did not have sufficient skills especially in script usage. Though they were quite capable of working with "paper and pencil" in their theoretical or experimental laboratories, their results always include human mistake. It is not due to their fault, but to the unavoidable fact from 'Statistics' that it is impossible to do everything manually without mistakes, occurring at the editing of input files, adjusting calculation conditions, and summarizing results. The use of scripts with text processing (awk/grep/sed) is powerful not only to avoid mistakes but also for finding mistakes quickly. By using 'grep' and 'awk' effectively applied to the input/output files, proficient collaborators can find such mistakes made by beginners quite quickly.

---

[18] In practical projects, there are also the cases to evaluate and compare the ground state energies for a fixed geometry, typically the experimental one [7].

4.6.2   Importance to Understand the Dependence on the Computational Conditions

In this chapter, we explained several dependences of the prediction on the conditions. The first half of the dependences mentions about the resolutions (E_{cut} and k-mesh) those are easy to understand as a procedure to look at the convergence. On the other hand, the topics in the latter half (choice of pseudopotentials and exchange-correlation potentials) are always annoying for beginners. Naive and direct question would be 'After all, which one should be chosen?', but no definite answer is obtained from experts, just getting vague comments such as 'Umm, it's quite difficult to answer in general...'.

As we will share this feeling in the next chapter, the choice of pseudo and exchange-correlation potentials gets too fundamental when one investigates deeper, seeming as an endless topic. A recommended attitude for practitioners would be just aware of the existence of the dependence on the choice, and

- When reading preceding papers related with your study, pay attention to the computational conditions explained in this chapter.
- When you work on your own calculations, be aware that the selection of exchange-correlation potentials is a difficult issue, being safer just to follow the same choice as preceding works took, or consult an expert.

Figure 4.3 shows an appealing example that the predictions differ qualitatively depending on the choice of exchange-correlation potentials [11]. It shows the prediction for C_{20} to take which structure is the most stable, shown in terms of the ground state energies for each possible polymorphs (Ring, Cage, or Bowl). 'HF', 'LDA', 'GGA', and 'DMC' shown inside the plot corresponds to the choices of exchange-correlation potentials.[19] The plot shows that each choice gives different prediction about which structure is the most stable.

Within the range of DFT, nobody cannot show any definite direction to upgrade the choice of exchange-correlation potentials for improved reliability of predictions (no systematic improvements), as we explained later in the next chapter. For this purpose, we have to refer the prediction obtained by another method, which is considered to be more reliable than DFT. Comparing the deviations between the reliable method and DFTs with different exchange-correlation choices, that giving the smallest deviation would be regarded as the proper choice. Such a study is called a **calibration of the exchange-correlation functionals**.

One might think, "If there is such a reliable method, why don't we use it?". However, the reliable method is like "a rescue car of the road assistance service", which is surely reliable but expensive. Furthermore, such methods can surely provide reliable ground state energy, but sometimes not practically feasible to provide other

---

[19] Exactly saying, 'HF' and 'DMC' are actually not the exchange-correlation potentials, but the name of other methods than DFT which handles the exchange and correlation in other framework.

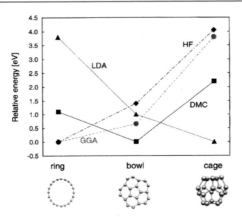

Fig. 4.3 Predictions for C_{20} to take which structure is the most stable, shown in terms of the ground state energies for each possible polymorphs (Ring, Cage, or Bowl) (figure taken from [11]). 'HF', 'LDA', 'GGA', and 'DMC' indicate the method for the predictions, corresponding to different ways how to handle the exchange and correlation of interacting electrons. Depending on the choice of methods, the most stable structure to be predicted differ from each other

properties like atomic forces etc. those DFT can provide. The methods are practically in limited use for the calibration for now. Incidentally, the author's group specialized in one of such method, quantum Monte Carlo (especially diffusion Monte Carlo; DMC as appeared in Fig. 4.3) electronic structure calculations (Sect. 6.3.3) [12].

Figure 4.4 shows a typical example of the calibration by DMC compared to DFT with several choices of exchange-correlation potentials [13]. In this target system, the binding is mainly formed by inter-molecular interactions, for which the conventional implementations of exchange-correlation potentials are not good at to capture the binding nature. The right panel of the figure shows positive or negative values of the binding energy depending on the choice, where negative values mean that the binding is properly reproduced. We can see at the left side in the right panel that the

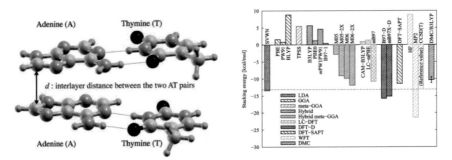

Fig. 4.4 The binding (along vertical) between the DNA bases evaluated by DFT with various exchange-correlation potentials and compared to reference values by DMC and CCSD(T) [13]. The right panel shows each binding energy. The negative values correspond to the reproduction of binding. Some methods are found not capable to reproduce the binding, giving positive binding energies

conventional choices like PBE or B3LYP are giving the positive, being deficient to reproduce the binding. These conventional ones have been developed for describing the covalent bonding first, and hence not took care of the inter-molecular ones.

The ones giving the proper negative values such as M05, ωB97, etc. belong to the younger generation, where the exchange-correlation potentials have been developed so that the deficiency of the conventional ones could be revised. Even in such younger ones, we find that some of them still fail to get negative but positive values. At the most right side in the right panel, we can see 'DMC' and 'CCSD(T)', that are the reference values for the calibration, evaluated individually by each reliable method. How close to this reference value can be a measure for an exchange-correlation potential to be justified for use in the system.

Judging from the coincidence alone, one might wonder that SVWN would be a good choice for this system because it gives quite closer value to the reference. It is, however, the consensus among experts that SVWN is never a proper choice for this system. We don't explain the reason here, but this is a sort of an *entry point* of the field ot experts in this field.

As in Fig. 4.4, it becomes the usual practice for DFT studies that not only a single choice but also various choices should be performed and compared in order to confirm the justification of a choice, since the computational cost required for a single choice gets enough cheaper.

References

1. "Adhesion of Electrodes on Diamond (111) Surface: A DFT Study", T. Ichibha, K. Hongo, I. Motochi, N.W. Makau, G.O. Amolo, R. Maezono, Diam. Relat. Mater. 81, 168 (2018). https://doi.org/10.1016/j.diamond.2017.12.008
2. "Molecular Electronic-Structure Theory", Trygve Helgaker, Poul Jorgensen, Jeppe Olsen, Wiley (2013/2/18), ISBN-13:978-1118531471
3. "Berry Phases in Electronic Structure Theory: Electric Polarization, Orbital Magnetization and Topological Insulators", David Vanderbilt, Cambridge University Press (2018/11/1), ISBN-13:978-1107157651
4. "Solid State Physics", N.W. Ashcroft and N.D. Mermin, Thomson Learning (1976).
5. "Linux Command Line and Shell Scripting Bible", Richard Blum, Christine Bresnahan, Wiley (2021/1/13) ISBN-13:978-1119700913
6. https://www.quantum-espresso.org (URL confirmed on 2022.11)
7. "Electronic Structure: Basic Theory and Practical Methods", Richard M. Martin, Cambridge University Press (2004/4/8), ISBN-13:978-0521782852
8. "GaN band-gap bias caused by semi-core treatment in pseudopotentials analyzed by the diffusion Monte Carlo method", Y Nikaido, T. Ichibha, K. Nakano, K. Hongo, and R. Maezono, AIP Adv. 11, 025225 (2021). https://doi.org/10.1063/5.0035047
9. Peter J. Heaney, "Structure and chemistry of the low-pressure silica polymorphs", Reviews in Mineralogy and Geochemistry 29 (1): 1-40 (1994) https://doi.org/10.1515/9781501509698.
10. "Quantum Theory of the Electron Liquid", Gabriele Giuliani, Giovanni Vignale, Cambridge University Press (2005/3/31) ISBN-13:978-0521821124
11. "Structure and Stability of Molecular Carbon: Importance of Electron Correlation", Jeffrey C. Grossman, Lubos Mitas, and Krishnan Raghavachari https://doi.org/10.1103/PhysRevLett.75.3870

12. "Quantum Monte Carlo simulations of solids" W.M.C. Foulkes, L. Mitas, R. J. Needs, and G. Rajagopal Rev. Mod. Phys. 73, 33 (2001). https://doi.org/10.1103/RevModPhys.73.33
13. "The Importance of Electron Correlation on Stacking Interaction of Adenine-Thymine Base-Pair Step in B-DNA: A Quantum Monte Carlo Study", K. Hongo N.T. Cuong, and R. Maezono, J. Chem. Theory Comput. 9, 1081 (2013). https://doi.org/10.1021/ct301065f

Points to Understand in Background Theories

5

Abstract

In this chapter, we explain the background theories as below for the operations that we experienced in preceding chapters. Readers are expected to harden one's understandings for each operation, confirming the background meaning for each procedure experienced before. "A school trip to visit historical heritages" can be used as a metaphor. Such a school trip would be more efficient rather for matured adults because they can harden their impressions on what they *experienced*. Background explanations *before* one experiences are sometimes not well working.

Items to learn

- **Positioning of the kernel calculation**
  We explain the relationship between kernel calculation and various evaluations for practical quantities, reaffirming our understanding that kernel calculation is the keystone to be learned.
- **Points to understand in kernel calculation**
  Among a lot of parameters to be specified in input files, what are the parameters to be understood at the first place.
- **Concept of density functional theory**
  We introduce the concept of "equivalent one-body form" which effectively represents the interacting many-body system. As the effective one-body potential, the exchange–correlation potential is introduced. We then explain the variations for the potential practically used.
- **Self-consistent iteration**
  We will understand what iterations are going around in the simulation to converge on what.
- **Pseudopotential**
  To what extent one should understand even at the user level, beyond which a user can leave the knowledge untouched as the expert's details.

© The Author(s), under exclusive license to Springer Nature Singapore Pte Ltd. 2023   121
R. Maezono, *Ab initio Calculation Tutorial*,
https://doi.org/10.1007/978-981-99-0919-3_5

- **Expanding basis functions as a factor for package choice**
  We learn that the package implementations have variations in the form of basis function expansions. We explain what ideas have led to the development of newer expansion forms recently. Based on this knowledge, we explain how to choose a package for your purpose in practice.

5.1 Significance to Learn Kernel Calculation

5.1.1 Kernel Calculation as a Seed

We have explained how to perform a reliable kernel calculation, namely, the evaluation of the electronic structures for a fixed lattice configuration, sometimes called **single-point calculation**. For the SCF iterative convergence in the kernel calculation, we used a metaphor described as "maturing dough". What is obtained by the maturing is the correct **electronic state** for a given system, which is a working "dough" to fabricate various bread products corresponding to the evaluations of several properties of the system.

The most fundamental example to utilize "the dough" is to form up informations from the band dispersion. The dispersion provides how dense the possible initial and the final states regarding the excitation to be considered. The visualization of this information corresponds to the spectrum evaluation, which is performed by calculation of the summation which appeared in Fermi golden rule. Based on the summation, it is possible to evaluate the dielectric constants [1], those functionality is available in any ab initio packages recently.

As another output of converged SCF (matured dough), we get a set of the orbital functions describing the electronic state. By using the set, further postprocessing calculations can be performed, such as **GW** calculations [2,28] (for more reliable estimation of the bandgap) or **QMC** electronic structure calculations (for more reliable ground state energy evaluations) [3]. These calculations are usually performed by individual packages beside the DFT packages. Users have to use some interfaces to extract the orbital functions, etc. from the output of DFT, transformed and written out to a file with different formats used in these different methods (GW, QMC, …). This is another example how the kernel calculations are used for further processing (like "bread products").

As explained in Sect. 4.3.1, the script works with loop structure can generate a set of input files automatically with slightly varying the geometry. For a uniform variation about the lattice constant, we can get the **bulk modulus** by fitting the plot of energy variation as the expanding coefficient of the second order with respect to the atomic displacement. For several choices how to give the displacements, we can get a set of **elastic constants** in the similar way [22].

When the displacements are given along the normal mode vibrations (Sect. 11.1.2), the corresponding "elastic constants" are nothing but the **phonon frequencies** [1].

There are several formulas empirically or model-theoretically developed to evaluate material properties (e.g., T_c for superconducting) using phonon frequencies. Recent DFT packages [6] provide such utilities inside to evaluate such formulas for transport, conduction [4], and superconducting properties [5] as the postprocessing over the phonon calculations.

The script can be fabricated so that the input geometries can be randomized. If we were lucky enough to find the lower ground state energy than that of known crystal structure ever by the random trial, then the structure is a promising candidate of the **novel structure unknown ever** to be realized. Though it is quite unlikely to find such structures for the ambient conditions, we can easily set the extreme conditions such as high pressure as in the mantle core expressed as the computational parameter setting [7]. Such **structure searching** was one of the source streams of Materials informatics to be developed further beyond the random search to **Bayesian** search [8] or that using **machine learning** [5,9].

For given tiny displacements (δa) of atomic positions deviating from each equilibrium, we can calculate how much the energy increased (δE). The gradient $\delta E/\delta a$ corresponds to the **force** originally defined by the derivative of E with respect to the displacement.

Based on such an idea, the force acting on each atom is evaluated numerically, driving each atomic position to perform ab initio **molecular dynamics** [10,22,23].[1]

5.1.2 Geometrical Optimization

Though it is not explained as a tutorial example since the present course concentrates on the kernel calculation, **the geometrical optimization** [23] would be an important task in a practical flow of research, commonly appearing even for beginner's work. The optimization is understood as another form of external script works[2] evaluating the force on each atom to update the positions until the force gets to be zero to arrive at the **equilibrium geometry** predicted by the simulation.

Assisted by recent computational power, the geometry optimization can be performed with enough faster speed, getting commonly used for practitioners. It can be a "*quick check*" (Sect. 3.7) for a choice of computational conditions seeing if the optimized geometries give the lattice constants closer to experimental ones (if available). In most of practical projects, it is usual to perform the geometrical optimization first to get the equilibrium geometry, and then the properties are evaluated [11]. If

---

[1] In practical implementations, the derivatives appeared in the molecular dynamics or the phonon calculation are actually evaluated not by such a *finite displacement method* but by more efficient algorithm. Nevertheless, it can be left as an advanced topic, and for the time being, the rough understanding of the concept is enough to be made by the finite displacement picture at the beginner's level.

[2] This is just a conceptual explanation. Practical implementation is more sophisticated and complicated being not the external script.

the geometry of a system to be considered is unknown, the geometrical optimization becomes a very powerful tool to *predict* the structure by the simulation.

Note that there are still the cases where we adopt a fixed geometry (usually experimental one) without any geometry optimization commonly applied to several computational conditions to be compared, depending on the purpose of one's analysis [12]. Some beginners are found to misunderstand that the geometrical optimization is the procedure that must be performed before everything, but it is not true. It is never the *all time procedure*. It is important to be aware of what geometrical optimization is for.

5.1.3   Why Kernel Calculations with Command Line

Figure 5.1 shows the relation between the kernel calculation and the secondary evaluations for properties. With this relationship in mind, one can understand why we have concentrated on leaning kernel calculations. Without understanding this, one might misunderstand that the "DOS calculation" and the "spectrum calculation" are in a parallel relationship. This is the reason why some beginners wander around without being able to correctly determine what they need to learn, saying "I want to do spectral calculations, how can I learn spectral calculations instead of DOS calculations?".

To use the metaphor of breads, regardless of what one wants to make, danishes or curry bread, what one has to learn first of all is how to make a properly matured dough. This corresponds to the kernel calculation, as the common foundation, for which one has to learn as an apprentice under the proper supervision because most of the new concepts for beginners are introduced at this stage. On the other hand, once one gets past this point, one can read the manual all by oneself to perform secondary calculations using "dough" to get several properties.

In Sect. 1.5.2, we mentioned that the GUI operation using "the mouse to click a button" is actually the execution of an equivalent command behind the scenes. In that section, we also discussed the relationship between GUI usage and command usage in electronic state calculation packages. The GUI usage of the packages is actually the vendor-provided "built-in way of handling" in the arrangement relationship shown in Fig. 5.1. Namely, a user cannot freely modify the way how to connect the results of kernel calculation to any secondary calculations arbitrary. Taking

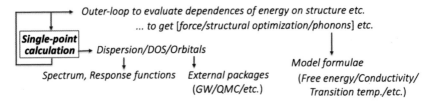

Fig. 5.1 While DOS and dispersion calculations are the direct output of the single-point kernel calculation, other calculations to evaluate several properties are the secondary ones using the converged result obtained by the primary kernel calculation

"MaterialsStudio" [13] as an example, it is a package integrating outer handlings over its kernel called "CASTEP" [14]. Most of practitioners would use is with GUI, where the package provides only popular, typical, and well-established handlings of secondary calculations connected from the kernel calculation as the "built-in" way. On the other hand, the researchers developing methodologies own CASTEP operated by the command line. By the command-line operations, they can further develop any new scheme for property evaluation composed of the script work of command line, freely arranging the connections in Fig. 5.1. As another example, "PhonoPy" [15] is a popular package to perform phonon property evaluations. This is a sort of outer script works over the existing kernel part, for which one can use CASTEP [14] or VASP [16] as a user specifies.

While the GUI usage is "built-in"-wise, the command-line usage of the kernel has further freedom. By learning the command-line handling as well as the script work, one can imagine how the package works to evaluate several properties as shown in Fig. 5.1. Another justification for learning the command line is related to the usage of supercomputers. When practical projects get to larger scale, one might be forced to use supercomputing facilities. Since GUI usage of a software is never common on such facilities, knowledge about the command-line usage is inevitably required.

5.2 Tips to Understand Kernel Calculation

5.2.1 Categorizing Input Parameters

The first goal for the beginners in this book is to be able to edit the input parameters by themselves as mentioned in Sect. 1.3.1. There are, however, so many input parameters involved in simulation packages that it is unrealistic to try to understand them from one end to the other in flat-list-wise. The package developers also understand that this is the most discouraging factor for beginners. It is then getting common practice for a package development to reduce the number of adjustable input parameters by hiding them behind with fixing default values to make invisible as the package developed. The ultimate example is the GUI usage of the package using "built-in choice" such as "SuperFine/Fine/Normal". In this way, the users can run simulations without being aware of how to set up proper input parameters [13].

In command-line usage, a user can freely adjust the parameters. The meaning of each parameter is usually explained in the web links that can be found on the web page of each individual package. The important tips for getting use of parameters are to **categorize them** into three groups rather than in a flat-list manner.

The **first group** corresponds to the parameters that can be understood without any particular expertise in electronic state calculations, such as the number of iterations, the threshold for convergence, the directory location for input and output files, etc. The **second group** of parameters is the "too detailed" level that can be left for the time being for the "user" level, not developer level. This category includes those such as specifying the convergence acceleration scheme, specifying the smearing scheme, specifying the initial guess for the charge distribution $n_0(\mathbf{r})$, etc. The latter two will

1st-group ; "*Parameters that can be understood even without any particular expertise*"
             e.g.) Num. of iterations, Threshold for convergence, Directory location of I/O files etc.
2nd-group ; "*Parameters to be learned later as they are advanced for the time being*"
             e.g.) Acceleration scheme, Smearing scheme, Initial guess for $n_0(r)$ etc.
3rd-group ; "*Parameters that beginners should learn first*"
             e.g.) exchange-correlation, pseudo potentials, Basis set functions.

                    Parameters that affect the reliability of the simulation predication.
                    Specs that must be included in the original paper.

Fig. 5.2 Rather than flat-list-wise, categorizing the input parameters into three levels would help users to understand and get use of them. The priority should be put for the **third** group to be understood first

be explained later in this book just briefly. Anyway, this "advanced" category is the matter to be understood afterward by reading more methodological textbooks after this book [17] (Fig. 5.2).

5.2.2   Input Parameters to Be Understood First

The **third group** is the parameters corresponding to the important concepts to be introduced first for the beginners. That includes "1/exchange–correlation potential", "2/pseudopotential", and "3/basis set expansion". These "BIG3" critically affect the reliability of simulation predictions, and should clearly be reported in any paper publications. Understanding these BIG3 concepts can be said as an important milestone that characterizes graduation from beginner level.

From the next section, we will learn the concepts about "BIG3" one by one in order.

5.3   Overview of Density Functional Theory: For Understanding Exchange–Correlation Potentials

5.3.1   Formulation of Our Problem

The problem addressed by ab initio electronic structure calculations in materials science is to determine the state of many electrons in motion in a material under the interaction from nuclei and other electrons. The possible states for N-electron system are all described by the **many-body wave function**, $\Psi\,(\mathbf{r}_1, \ldots, \mathbf{r}_N)$, which is the eigenfunctions determined by the **many-body Schrödinger equation**,

$$\left[-\frac{1}{2} \sum_{j=1}^{N} \nabla_j^2 + V\,(\mathbf{r}_1, \ldots, \mathbf{r}_N) \right] \cdot \Psi\,(\mathbf{r}_1, \ldots, \mathbf{r}_N) = E \cdot \Psi\,(\mathbf{r}_1, \ldots, \mathbf{r}_N) , \qquad (5.1)$$

as given in the framework of the many-body wave function theory [18]. When a user specifies the atomic geometry (the spatial configuration of atomic nuclei) of the system, the functional form of the **potential term**, $V\,(\mathbf{r}_1, \ldots, \mathbf{r}_N)$, is fixed to describe the interactions between particles (electrons and nuclei). By this way, the partial differential equation to be solved is defined.

What is meant by "solving the above equation" is to find the possible combinations of (E, Ψ) that satisfies the equal relation called as eigenvalue problem of differential equations [19]. Since there are several pairs of (E, Ψ) that satisfy the above, we label them with j as $\{E_j, \Psi_j\}_{j=0}^{\infty}$. The corresponding $\Psi_0, \Psi_1, \ldots$ are the **electronic states** at each energy level, $E_0, E_1, \ldots$. The clarification of possible energy values and electronic states is the problem to be solved by the many-body wave function theory.

The many-body electronic state $\Psi(\mathbf{r}_1, \ldots, \mathbf{r}_N)$ is like a deity that says, "If you integrate me accordingly, you can get various physical quantities". It may indeed contain all the information, but it is like an elusive entity that is difficult to be imagined. With the argument $(\mathbf{r}_1, \ldots, \mathbf{r}_N)$,[3] it forms complex values depending on them as a function in the high-dimensional $(3N)$ space.

> **! Attention**
>
> The pain point of many-body wave function theory is said that the description dimension gets higher depending on the number of particles. The descriptive dimension of a three-particle system is nine dimensions, while that of a four-particle system is 12 dimensions, but the phenomenon is commonly occurring in three-dimensional space. It is surely uncomfortable to have different descriptive spaces for each fundamental principle. This is especially true when dealing with phenomena with variable particle number such as grand canonical ensemble or quantum condensation. The framework using $3N$-dimensional description was therefore regarded as not the ultimate form of the theory, which motived the creation of quantum field theory that describes many-particle systems by the three-dimensional space argument [20].

Since multivariate functions are difficult to deal with, a conventional strategy is taken to **assume**, for example, $g(x, y) = f_1(x) \cdot f_2(y)$, as a product of lower dimensional functions, called **variable separation method** [19]. By this way, we can handle the problem in terms of the lower dimensional space. For our many-body wave function, it leads to

$$\Psi(\mathbf{r}_1, \ldots, \mathbf{r}_N) \sim \psi_1(\mathbf{r}_1) \psi_2(\mathbf{r}_2) \ldots \psi_N(\mathbf{r}_N) , \tag{5.2}$$

as an **assumption**. By this assumption, solving the equation given in $3N$-dimensional space reduces to the problem described by a set of three-dimensional functions $\psi(\mathbf{r})$. This form is called **one-body form**.

Without the interaction V,[4] the variable separation is justified, and "solving the original many-body Eq. (5.1)" is replaced by "solving a set of N one-body equations".

---

[3] It describes a snapshot of a configuration by N-electrons in three-dimensional space.

[4] To be precise, "without electron–electron interactions" contained in V.

Since these solutions $\psi_j(\mathbf{r})$ can be viewed as "distributions in three-dimensional space", we can get a picture of "electronic orbitals with such distributions", called **orbital function**.

The use of "~" instead of "=" in Eq. (5.2) is tentative, because we still need to improve the assumption for taking into account the following circumstances.

In order to obtain a solution describing the electrons properly, we have to solve the many-body Schrödinger equation under the constraint **anti-symmetric constraint** [10,30] that

$$\Psi\left(\ldots, \mathbf{r}_j, \ldots, \mathbf{r}_i, \ldots\right) = (-) \cdot \Psi\left(\ldots, \mathbf{r}_i, \ldots, \mathbf{r}_j, \ldots\right) . \tag{5.3}$$

This is the fermionic condition that when two particle coordinates are interchanged, the sign is reversed while preserving the absolute value, which leads to the Pauli exclusion rule. The assumption of Eq. (5.2) never satisfies the above anti-symmetric constraint. The most naive fabrication for satisfying the constraint is to take an anti-symmetrized linear combination,

$$\Psi_{SD}\left(\mathbf{r}_1, \mathbf{r}_2, \ldots, \mathbf{r}_N\right) = \sum_P (-)^P \cdot \psi_1\left(\mathbf{r}_{P_1}\right) \psi_2\left(\mathbf{r}_{P_2}\right) \ldots \psi_N\left(\mathbf{r}_{P_N}\right)$$

$$= \begin{vmatrix} \psi_1(\mathbf{r}_1) & \cdots & \psi_1(\mathbf{r}_N) \\ \vdots & \ddots & \\ \psi_N(\mathbf{r}_1) & \cdots & \psi_N(\mathbf{r}_N) \end{vmatrix}, \tag{5.4}$$

where we used the definition formula for a determinant. The above determinant is called **Slater determinant** [10,30].

While the assumption is justified when without interactions in V, it is not always ensured with finite interaction that the exact solution is expressed by the simple variable separation. In the general cases, the form of Slater determinant is just an approximation. The method adopting this approximation is the **Hartree–Fock** theory [10,30], where the orbital functions are optimized so that the approximation gets to be closest to the exact solution within the possible extent to be described by the form of Eq. (5.4). In other words, the Hartree–Fock approximation applies the form (variable separation and orbital picture) that is completely justified when the absence of electron–electron interactions to the case even with finite interactions [18]. The approximation hence takes into account the interaction itself, but it cannot take into account the effect of the functional deformation by the interaction deviating from the functional form without interactions. As the interaction gets stronger, the question arises, "Does the functional form justified for the non-interacting case really reflect the reality with stronger interaction?"[5] There are a variety of theories to incorporate to capture the **deformation effects of the many-body wave functions**, including

---

[5] This is one of the motivations for the research field called **strongly correlated systems** that aims at capturing the exotic situations beyond the possible extent to be described by one-body orbital picture.

the correction using multi-determinants or adopting functional forms completely different from the product of one-body orbital functions (e.g., geminal, Pfaffian, backflow, etc.) [21,47].

5.3.2 Introducing Density Functional Theory

In the usual view for the systems with non-weak interactions, one would consider how to take into account "the deformation from the one-body orbital form". Taking a sort of "inverted view", a question was posed, "Is there any **equivalent one-body form** that can give exactly the same answer as that described by the deformed one with non-weak interactions". In other words, it sets up the problem that "Is it guaranteed to map an interacting system of arbitrary strength into a **equivalent one-body form**?". To this question, a positive answer ("yes") with mathematical proof was given, called density functional theory [18].

The "equivalent answer" means that the same answer of the energy and charge density given by the exact solution is also given by the equivalent one-body form as ensured to exist. It is amazing that such a thing can be guaranteed to be exact, and for this work achieved in the 1960s, Prof. Walter Kohn was awarded the Nobel Prize in 1998. This framework is based on what is called the Hohenberg–Kohn theorem, which states that the ground state energy E_{GS} is determined by the charge density alone ($E_{GS} = E_{GS}[n]$). Originally, in order to get the exact value of E_{GS}, we have to solve the many-body Schrödinger equation which becomes quite hard task for interacting systems. The theorem, however, ensures the possibility to avoid such a hard task because we can get the same exact value of $E_{GS}[n]$ by getting the same exact charge density if it can be obtained by some alternative method with much reduced complexity (Fig. 5.3).

$$n(\vec{r}) = n\left[\Psi_{\text{interacting}}\left(\vec{r}_1, \cdots, \vec{r}_N\right)\right]$$

Exact solution (largely deformed from one-body form

$$n(\vec{r}) \longrightarrow E_{GS} = E_{GS}\left[n(\vec{r})\right]$$

by interactions)

$$n(\vec{r}) = n\left[\Psi_{\text{one-body prob.}}\left(\vec{r}_1, \cdots, \vec{r}_N\right)\right]$$

... doesn't matter how to get. $n(\vec{r})$ ...

charge density formed by an equivalent one-body form

Fig. 5.3 In DFT, it is proved that the ground energy E_{GS} is determined by the charge density alone ($E_{GS} = E_{GS}[n]$). If we can use another form that is easier to solve and gives the same $n(\mathbf{r})$ as the exact solution, we can get exact E_{GS} as ensured. Even without knowing how to get the exact solution, how to find an equivalent form giving the same solution is a next big question. However, a series of procedures based on variational forms have been established to achieve that in DFT

Tea Break

While the research called "strongly correlated systems" [2] has been gaining momentum with the fascinating phrase "a mysterious and unexplored area where the familiar one-body formalism does not hold", the concept that "many-body problems can be solved exactly by equivalently mapped one-body forms" has not well been accepted in the research trend. It has been called "band calculations" for a long time, being lumped together with the one-body "approximation" approach before the advent of density functional theory. Many DFT experts have tried to correct it appearing "the existence of a mapping to the one-body form had mathematically proved (in the 1960s!)" even if the target is strongly correlated, but they could not overcome the overwhelming impression that the one-body approximation is not working, giving a novel field to be investigated. It is true that the existence of mappings is limited to the ground state of normal phase, and does not apply to problems such as excited states and phase transitions, which have been the main interest in strongly correlated systems. However, there has been a widespread misconception that the application of DFT is fundamentally wrong, even for the description of the ground state of normal phase if the system is regarded as strongly correlated.

The equivalent one-body form to give the same charge density (called **reference system**) as that of the exact solution takes the form[6] [4,23],

$$\left[-\frac{1}{2}\nabla^2 + v_{\text{ext}}\left(\mathbf{r}\right) + \hat{U}_{\text{ele.}}^{(\text{eff})} \right] \cdot \psi_j\left(\mathbf{r}\right) = \varepsilon_j \cdot \psi_j\left(\mathbf{r}\right) \quad (j = 1, \ldots, N) . \quad (5.5)$$

Once we have obtained the set of solutions for this eigenvalue problem, $\left\{ \varepsilon_j, \psi_j(\mathbf{r}) \right\}_{j=1}^{N}$, the exact ground state energy and the charge density are given as

$$E_G = \sum_{j=1}^{N} \varepsilon_j, \quad n\left(\mathbf{r}\right) = \sum_{j=1}^{N} \left| \psi_j\left(\mathbf{r}\right) \right|^2 , \quad (5.6)$$

as derived in the Kohn–Sham framework [22,23].

The terms, $v_{\text{ext}}\left(\mathbf{r}\right)$ and $\hat{U}_{\text{ele.}}^{\text{eff}}$, appearing in Eq. (5.5), are both the potential felt by a single electron (the jth electron). The $v_{\text{ext}}\left(\mathbf{r}\right)$ is the potential felt by an electron at $\mathbf{r}$ from the ionic core of each element specified as the geometry. The $\hat{U}_{\text{ele.}}^{\text{eff}}$ is the "effective" one-body representation of the electron–electron interaction from the other electrons. This term is in response to the question, "Is it really possible for the many-body interaction with arbitrary strength to be mapped to one-body form?". It

---

[6] Note the difference from Eq. (5.1). In Eq. (5.5), ψ is lower case and its argument is the three-dimensional coordinate $\mathbf{r}$. On the other hand, in Eq. (5.1), Ψ is uppercase and its argument is a snapshot of the positions of N-electrons $(\mathbf{r}_1, \ldots, \mathbf{r}_N)$.

has a meaningful caret (^) on it which implies that the term cannot be expressed by a simple function.

Decomposing $\hat{U}_{\text{ele.}}^{\text{eff}}$ as

$$\hat{U}_{\text{ele.}}^{(\text{eff})} = u_{\text{class}}(\mathbf{r}) + \hat{u}_{XC} , \tag{5.7}$$

the contribution that can be written as a classical electrostatic interaction is separated as $u_{\text{class}}(\mathbf{r})$. This part is called the **Hartree term** [22,23], which is the sum of the electrostatic potential due to the charge density created by the one-body orbitals of other electrons, $|\psi_i(\mathbf{r})|^2$. The remainder of $u_{\text{class}}(\mathbf{r})$ is the "effective potential representing the quantum effect beyond the classical one", written with the caret $\hat{u}_{XC}$.

Tea Break

As mentioned, DFT was not so well accepted in the community of condensed matter physics in Japan at least up to around 2000s. There was a similar atmosphere in the community of quantum chemistry at that time. The reason for the latter community is a bit different from that in the former. They are a bit allergic against such a "mapping theory" where one cannot estimate well how descriptiveness is improved by modifying the formalism (not like molecular orbital theory where the systematic improvement is assured to some extent). When a big influencer mentioned "I don't like DFT. It's never 'ab initio!'", the atmosphere of the community makes it difficult to mention "it ab initio study by DFT". As another story about "ab initio", the relation with the community of elementary particle physics was also interesting, which is involved in the grand community around the national flagship supercomputer project. Sometimes they complained that "Taking the mass of an electron as a given is never 'ab initio'…". As such, speakers talking about DFT contents had to say something like "I am sorry, but please forgive me if the word 'ab initio' comes out of my mouth during the lecture, but it's just a custom in our community…"

"XC" is an abbreviation for "Exchange–Correlation", meaning **exchange effect** and **correlation effect**. In Sect. 5.3.1, we explained the Hartree–Fock method, which takes into account the interaction itself but not the effect of the deformation of wave function made by the interaction. This deformation effect is called "correlation effect" in the traditional wording. The exchange effect is that reflecting the Pauli exclusion law as expressed by the anti-symmetric constraint, being an effect beyond the classical formalism. In Hartree–Fock theory, the exchange effect is taken into account as imposed by Slater determinant, while the correlation effect is not.

Tea Break

The correlation effect is explained above with the term "in the traditional sense". The electronic correlation has traditionally been defined as the "effect that cannot be captured by Hartree–Fock theory", and there has been a long accumulation of

Meanfield	+ ***Fluctuations*** *(electronic correlation)*
= [*Classical meanfield*] +	[(*Quantum statistics* (***exchange***))	+ (*Electronic **correlation**)*]
= [*Classical meanfield*]	+ [***Exchange-correlation*** *potentials*]

$$** \text{ Quantum statistics ... Effects due to Pauli exclusion rule.}$$
$$** \text{ Electronic correlation ... Effects of deformation in wavefunction by interactions}$$

Fig. 5.4 A distinction on the effect of electron interaction. The word "correlation" tends to be abused in various contexts, but it is important to understand the above distinction

research under this definition. As the research topics on strongly correlated systems gained momentum after high T_c superconductivity, the term "electronic correlation" has sometimes been abused in various contexts, mainly because of its wordy impression, and often conflicts with the above "traditional definition". It may seem natural for non-experts to describe strong electron–electron interaction as "strong electron correlation", but often the resulting large **exchange** effect is described as "strong correlation effect as conflicting the traditional meaning.

We would like to give a little more attention to one point. In Eq. (5.5), each term looks as if it were "this term is the kinetic energy [nabla term], and this term the classical electrostatic interaction", but it should be noted that these are "for ψ_j in the reference system". For the real problem realized as physics, these are not necessarily being the kinetic and classical electrostatic interaction, respectively. Specifically, the "kinetic energy value (in the reference system)" evaluated by the DFT calculation may include the energy due to the interaction in the real system (Fig. 5.4).

5.3.3   Exchange–Correlation Potentials

The $\hat{u}_{XC}$ appearing in Eq. (5.7) is called the **exchange–correlation potential** [1, 10, 18, 22, 23]. The tricky thing about it is that only its existence is assured. Although it is assured that there is an exchange–correlation potential that reproduces the exact charge density, we do not know how to create it, and we have to approximate it based on various possible guidelines as explained later in Sect. 5.3.6. This leads to the variation of exchange–correlation potentials available for a user. Depending on the choice of exchange–correlation potentials, one gets different predictions as seen in Sect. 4.5.

Knowing that "only the existence is assured but not how to build it up", one might have disappointing impression that "nothing is solved after all". However, this existence proof is a great thing because it guarantees to obtain exact ground state solution of many-body interacting system just by focusing on improving the exchange–correlation potential. It's a great relief to be ensured for us "pay attention just here!", no need to look around for "fires break out from anywhere else".

One may have another impression that "even though the framework is rigorous, practical implementation is an approximation after all". However, what is often misunderstood here is that the meaning of "approximation" is slightly different from that of "approximation in model theory". In the latter, the approximations are made in the positive sense to ignore unessential parts with physical perspective. On the other hand, approximations in exchange–correlation potentials are not like that, but rather on arithmetic considerations, such as, "Let's drop this for the time being, because it would be too costly to evaluate it numerically". Approximations in the exchange–correlation potentials is the issue on the **mathematically mapped** reference systems. It is hence impossible in principle for an approximating operation in mapped system to correspond to which meaning physically in a realistic system.[7]

Up to a certain level, the exchange–correlation potential is universally determined only by the kind of **bare** inter-particle interaction (Coulomb or nuclear, etc.) ($\rightarrow$ Sect. 12.3). It is important to understand that this is an approximation made for arithmetic reasons for mathematical entities that do not have much to do with the physical picture, such as covalent, metallic, or intermolecular bonding. Some people ask, "You're still using some kind of 'bribe' to fit the results, aren't you?", but this is a misunderstanding due to the confusion between approximations in ab initio framework and those in model theory (subjective approximations to physical descriptions).

5.3.4 Kohn–Sham Equation

What we are calculating to solve inside the package is a series of equations, Eq. (5.5), which is called the **Kohn–Sham equation** [1, 10, 18, 22, 23]. The equations describe the satisfaction that the functions $\{\psi_j\}$ obtained as the solution can form the exact charge density $n(\mathbf{r})$ as in the form as Eq. (5.6). For the question "What is the meaning of the orbital function $\{\psi_j\}$?", the answer is quite formal: "It is the functions introduced so that its sum of squares expresses the exact charge density". The orbital function introduced in this way is called **Kohn–Sham orbitals**. The Hartree–Fock orbitals described in Eq. (5.2) are in a good contrast: they are introduced to perform the variable separation for original many-body wave function.

Since both the Hartree–Fock and Kohn–Sham equations take the same form in their structure,[8] they can be implemented and realized on the same package.[9] However, it is important to note again that the Hartree–Fock orbital and the Kohn–Sham orbital are introduced in quite different contexts, as described above. In the most stoic viewpoint, one could say that the Kohn–Sham orbital has no physical meaning beyond the mathematical context (namely, they are introduced just to describe $n(\mathbf{r})$ as their sum of squared quantities). Nevertheless, if one regards the Hartree–

---

[7] This might be a weakness of DFT: when an approximation is found to be inadequate, it is not systematically clear what improvements can be made to improve the descriptiveness of the physics.

[8] The Kohn–Sham formalism is actually designed so that it may take the same form as that of Hartree–Fock theory in order that the same picture may work to understand the results.

[9] This is the case for packages such as "Gaussian" used for isolated molecular systems. For periodic systems, the Hartree–Fock calculation is usually not possible because the evaluation of the exchange term gets costly for the infinite reputation in a periodic system.

Fock method as a "DFT with zero correlation terms",[10] one would be possible to say that the Kohn–Sham orbital and the Hartree–Fock orbital are connected in this limit, where the Hartree–Fock orbital has physical meaning/image to some extent. The "connection" would justify indirectly such efforts to discuss the shape and the energy level of the Kohn–Sham orbitals as if they have the physical reality like molecular orbitals by visualizing them in many practical studies.

Note that the term **"density functional** theory" claims that the ground state of a system can be described as a functional of the charge density $n(\mathbf{r})$.[11] When we denote a **functional** $F[n]$, it means that $F[n]$ depends not only on the function $n(\mathbf{r})$ itself, but also on derivatives $\nabla n(\mathbf{r})$, $\Delta n(\mathbf{r})$, etc. [19]. In any case, $F[n]$ can be determined by giving $n(\mathbf{r})$. We have been attaching meaningful carets to $\hat{U}_{\text{ele.}}^{\text{eff}}$ and $\hat{u}_{\text{XC}}$ so far, but they are meant to imply that they are not just functions, but functionals.

5.3.5   Self-consistent Field Form

DFT leads to the fact that $\hat{U}_{\text{ele.}}^{\text{eff}}$ appearing in Eq. (5.5) is a functional of density as denoted $\hat{U}_{\text{ele.}}^{\text{eff}}[n]$. Then, we notice the following structure: If we want to solve Eq. (5.5), the charge density $n(\mathbf{r})$ should be given to fix the equation, as shown in Fig. 5.5. However, $n(\mathbf{r})$ cannot be determined unless we solve Eq. (5.5) in the first place. We fall into the cycle of "egg first or chicken first".

For this structure of the problem, we take the strategy of iterations: Solving the equation first with the initial guess of the charge density $n_0(\mathbf{r})$, we can get the resultant charge density $n_1(\mathbf{r})$. Using the updated $n_1(\mathbf{r})$, we solve the equation to get $n_2(\mathbf{r})$ ...If we get the convergence of $n_k(\mathbf{r})$ $(k = 2, \ldots)$ by repeating this process,

$$\left[-\frac{1}{2}\nabla^2 + v_{ext}(\vec{r}) + \hat{U}_{\text{ele.}}^{(eff)}[n] \right] \psi_j(\vec{r}) = \varepsilon_j \psi_j(\vec{r}) \qquad n(\vec{r}) = \sum_i |\psi_i(\vec{r})|^2$$

Fig. 5.5 Self-consistent (SCF) loop to solve the Kohn–Sham equation. To determine the equation specifically, the charge density $n(\mathbf{r})$ must be given, which is, however, the quantity obtained by solving the equation in the first place. Therefore, in practice, we solve the equation using the initial guess $n_0(\mathbf{r})$ to get the answer to compose the resultant $n_1(\mathbf{r})$. Using $n_1(\mathbf{r})$, solve the equation to get $n_2(\mathbf{r})$ ... Repeating the iteration to get the series $n_k(\mathbf{r})$ $(k = 2, \ldots)$, the converged result is the solution of the equation

---

[10] Correctly speaking, it is "exact exchange DFT".

[11] Recalling that the charge density is a quantity determined by the absolute value of the wave function, the statement, "...determined only by the charge density", implies that the phase angle of the wave function does not matter. It might be surprising because we have taught that the wave function being a complex function with a phase is the essence of quantum mechanics. Of course, we realize that this statement is limited to the ground energy of the system, but still, it is a bit of a surprise.

the converged one is the solution for $n(\mathbf{r})$. This iteration is called "self-consistent field (SCF) calculation" [1,10,22,23,30].

The convergence is not always achieved smoothly in a SCF loop. Such a situation was mentioned before in the second paragraph of Sect. 4.1.1. In Appendix Sect. 10.2, we explained the examples of poor convergence and how to improve the convergence. The choice of the initial guess $n_0(\mathbf{r})$ also matters much for the convergence. Any package provides a default setting of $n_0(\mathbf{r})$ automatically [just as the superposition of each isolated atomic density], but the default guess sometimes does not work well for the convergence. In such cases, proficient users have the skill to tune $n_0(\mathbf{r})$ to be more plausible for a given system, and are able to successfully bring about convergence. Such skill can be said as a form of intangible knowledge that is passed on in skilled research groups.

5.3.6 Variation of Exchange–Correlation Functionals

Though the existence of the exact exchange–correlation functional to reproduce the exact $n(\mathbf{r})$ is ensured, the concrete form or the procedure to build the functional is never provided, as explained in Sect. 5.3.3. Therefore, there have been many efforts to build it in trial-wise. For the variety of the available exchange–correlation functionals, an user has to specify which one to use in the input file. Depending on the choice, the achieved prediction varies significantly as explained in Sect. 4.6.

In Sect. 4.5, we tried several different functionals such as PZ/PBE/PW91/BLYP. Although these names may seem difficult to understand and remember, most of them are a combination of the first letter of developers and the year when it appeared, e.g., "PW91" means the functionals developed by Perdue and Wang in 1991.

Many functionals have been developed so far along various different guiding principles, such as those listed below:

- **a/Variation with respect to target systems:**
  - 1a/Periodic solids or isolated molecules.
  - 2a/Difference of the bonding nature, such as metallic, ionic, covalent, intermolecular, etc.
- **b/Variation with respect to guiding principles:**
  - 1b/Constructed to satisfy formal (mathematical) requirements.
  - 2b/Constructed to achieve the best fit for experimental values.
- **c/Variation with respect to improvement guidelines:**
  - 1c/Formal (mathematical) viewpoints (e.g., incorporating higher order derivatives).
  - 2c/Physical process viewpoints (e.g., incorporating intermolecular interactions).

5.3.7   Further Notes on Exchange–Correlation Potentials

The simple question from practitioners would be "After all, which exchange–correlation functional should I use?". As mentioned in Sect. 4.6.2, however, it is difficult to draw a clear conclusion to the question. Here, we explain a little further with some selected topics where the choice of the exchange–correlation potential requires careful attention. This is something that practitioners should be aware of when they submit a paper and receive reviewer comments such as "how to justify your selection of the exchange–correlation functionals?" For the following "difficult problems depending heavily on the selection of exchange–correlation potentials", the ab initio quantum Monte Carlo method [3] introduced in Sect. 6.3.3 is a powerful calibration method. This was also mentioned in Fig. 4.3 in Sect. 4.6.2.

5.3.7.1 Local Density Approximation

Since the quantum many-body problem has been studied in detail for homogeneous electron gas, exchange–correlation functionals have also been developed based on the knowledge on homogeneous electron gas.

One may be wondering why we can use the findings of the homogeneous electron gas to describe a real/inhomogeneous system with nuclei array. As mentioned with keyword "universal functional" in the footnote at the end of Sect. 5.3.3, the exchange–correlation functional describes the interaction between quantum mechanical particles and has the property of being universally valid for any system if the bare interaction[12] is the Coulomb interaction (Sect. 12.3). Since the universality holds even for the idealized electron gas systems as well, we are justified in using the knowledge there. The situation has slightly different nuance from the simply saying "approximated by the homogeneous electron gas".

One might also get such impression that "It's too rude to apply the knowledge of homogeneous electron gas to the realistic systems with inhomogeneous distribution of electrons". However, we note that we never approximate the system as the entirely homogeneous system. As we did in the calculus course, any curve can be described as a bar graph as far as it is enough smooth. If the change in the electron density is smooth as a continuous function, then it can be treated as the array of tiny homogeneous electron gas systems locally around each position to be considered. Such treatment would be justified unless the system is non-analytic situation where the application of differential calculus is questionable. This treatment is called the **local density approximation (LDA)**. It is the longest established exchange–correlation functional, which appeared in Sect. 4.5 labeled as "LDA".

In this context, we can understand that the difficulty would generally appear when we treat surface/interface and local defect problems using conventional LDA-based

---

[12] The "bare interaction" means the interaction of the grand origin. As a result of the complex interplay in the bare interactions, the intermolecular interactions ($\sim 1/r^6$) or the shielded interactions ($\sim e^{-\alpha r}/r$) are derived as the form of effective interactions. They are, however, all coming from the bare Coulomb interactions $\sim 1/r$.

exchange–correlation functionals. These systems have abrupt charge density changes that are delta- or step-function-wise at the defect point or interface. For such situation, it gets a little more difficult to ensure that the charge density is smooth enough.[13]

Intermolecular interactions

The dispersion interaction [24] or van der Waals (vdW) interaction is a representative factor of intermolecular interaction, being known as a difficult process to be described by the conventional exchange–correlation functional [25]. The interaction is explained as an electrostatic coupling caused by the quantum fluctuation to produce the charge polarization at one charge source that induces opposite polarization at distant charges. It is a nonlocal phenomenon in which an event occurring at one location depends on an event occurring at a remote location. Such nonlocal phenomena is out of the scope of the local density approximation, in the first place. In Fig. 4.4, we mentioned that the local approximation, LDA (SVWN), "appears to describe" the intermolecular binding, but experts consider it spurious. The following is a brief explanation why such an apparent binding is concluded in LDA.

The reason is explained as the "exchange repulsion" is underestimated [26]. The Pauli exclusion prevents electrons from approaching each other, which is expressed as a repulsion force when the distance between electrons gets closer (exchange repulsion). In LDA, the exchange repulsion is not fully described due to the deficiency of the framework as described in Sect. 12.1. What would be unapproachable under the true exchange repulsion becomes more easily approachable due to the lack of full description of exchange repulsion. This is "underestimation of exchange repulsion" [26].

While the nonlocal intermolecular interaction is not well described in LDA to lead unreasonably weak attraction, the exchange repulsion is also predicted to be weaker due to the deficiency of approximation, getting the overall attractive binding. This is what is thought to be the reason for the spurious binding in LDA.

Imperfect cancellation of self-interaction causing gap problem

We stated that in LDA the exchange interaction gets to be reduced from that should originally be there. This problem is called **the imperfect cancellation of the self-interaction** (Sect. 12.1), as explained below. In Sect. 5.3.2, we introduced the notations with caret, and the concept of functional. $E_{GS} = E_{GS}[n]$ means a quantity identified by the charge density $n(\mathbf{r})$, but the notation implies that E_{GS} cannot be identified solely by the value of $n(\mathbf{r})$, but depends also on the higher order derivatives like $\nabla n(\mathbf{r})$, etc. In this context, LDA means the treatment to approximate the functional $E_{GS}[n]$ by the function $E_{GS}(n)$.

As explained in Chap. 12, it is actually impossible to represent the exchange interaction J as a function $J = J(n)$. Hence, LDA manages to approximate it as

---

[13] This difficulty is a different factor from that described in Sect. 6.1.1 about the computational cost caused by the larger unit cells.

a function, leading to the damage of the balanced relation between the exchange and correlation terms. Due to the balance, there is an exact cancellation of the self-interaction terms, each emerging from exchange and correlation interactions, respectively. When the balance is damaged, the exchange part gets to be underestimated. The underestimation leads to the bandgap underestimation [1], which is well-known deficiency of LDA. The mechanism for this is explained in detail in the separate appendix Chap. 12.

The damaged balance leads to the imperfect cancellation of the self-interaction terms. Sometimes, the situation is vaguely mentioned as "the problem of self-interaction", but this terming is a bit confusing, as an origin of the misunderstanding about the gap underestimation and DFT+U methods. The problem is not for "the self-interaction", but "the cancellation of the self-interaction". The problem, namely, the spurious reduction of the exchange part, gets to be pronounced when the charge density gets to be localized (Chap. 12). While the itinerant systems are not so seriously suffered from the problem, the systems with localized charge densities are affected more, as exhibiting the bandgap underestimations.

Magnetism under the balance between exchange and correlation

Magnetism is based on a delicate balance between exchange and correlation effects. One may have heard of Bloch's theory of ferromagnetism or Kanamori theory [27] in the theoretical course of magnetism. Ferromagnetism could be explained by the exchange effect at first glance (Bloch's theory), but such ferromagnetism disappears as an artifact when the correlation effect is taken into account. Then the ferromagnetism is explained by other factors (Kanamori theory). In further detailed theories,[14] the prediction is known to be critically depending on the model construction in the way of the balance between the exchange and correlation terms. In these models, the balance should be observed so that when the correlation is incorporated to this level of precision, the exchange should also be incorporated to the same level of precision to satisfy symmetry as physics.[15]

Exotic properties of materials such as magnetism or superconductivity are largely the result of delicate balance between the exchange and correlation. To describe such systems by DFT, a proper description is not possible unless the exchange–correlation functional is carefully constructed.[16]

---

[14] They are usually described by the diagrams appeared in the field theoretical perturbation theory.

[15] A professor once metaphorically said that any lie can be derived if the exchange and correlation terms are freely set up without regard to balance.

[16] The problem originates at the stage of the fabrication of exchange–correlation functionals, because we don't know how to fabricate the exact functional which is ensured for its existence even for the description of the magnetism. It is never the problem of DFT itself. There used be significant numbers of "big professors" who misunderstood this point, claiming that the magnetism is completely out of the scope of the DFT, which takes the one-body form to capture the difficult many-body problems.

DFT+U

Regarding to the bandgap problem, the DFT+U method gets to be popularly used recently [28]. Other than the DFT+U, there are several alternative ways (e.g., exact exchange), but the DFT+U gets to be the most commonly used as implemented in standard DFT packages. It has become a well-known method even for beginners when they try dealing with the compounds containing d and f electrons, as a required framework. The explanation often heard says that "d-electron systems have the electronic correlation problem which is properly described the Hubbard model, but the U parameter there cannot be described by DFT, so it is like putting U by hand". However, this understanding is a bit inaccurate. The essence of the matter is that "when the localization gets pronounced, the cancellation of the self-interaction gets damaged, so DFT+U is to compensate for that". It is not for d electrons, so electronic correlation. DFT+U is more a matter of the exchange effect than the correlation effect: When the self-interaction cancellation is damaged, what becomes inadequate is the expression for the exchange effect, not the correlation effect (the effect caused by distortion of the wave function). It has been a long time since a bit wrong understanding of the DFT+U gets populated. Recently, we often get such reviewer's comments that "since you are dealing with d-electrons, you should use DFT+U" even though we are dealing with non-localized system. Since this is a matter of increasing contact with practitioners, we will discuss this situation in more detail in the appendix chapter Chap. 12.

5.4 Basis Set Functions

5.4.1 Basis Set Expansion

Including the Kohn–Sham equation, the eigenvalue problems with a linear operator $\hat{L}$,

$$\hat{L} \cdot \psi\,(\mathbf{r}) = \varepsilon \cdot \psi\,(\mathbf{r})\ , \tag{5.8}$$

are usually treated by the expanded functional series [29],

$$\psi\,(\mathbf{r}) = \sum_{l=1}^{M} c_l \cdot \chi_l\,(\mathbf{r}), \tag{5.9}$$

where $\{\chi_l\}_{l=1}^{M}$ are the basis set functions to expand the solution. In this way, the function ψ can be regarded as a vector,

$$|\psi\rangle \sim (c_1, \ldots, c_M)^T\ , \tag{5.10}$$

and the eigenvalue problem in a differential equation form Eq. (5.8) can be represented by the matrix eigenvalue problem as follows: Substituting the expansion Eq. (5.9) into Eq. (5.8) and reverting the left- and the right-hand sides,

$$\varepsilon \sum_{l=1}^{M} c_l \cdot \chi_l \left(\mathbf{r} \right) = \sum_{l=1}^{M} c_l \cdot \hat{L} \chi_l \left(\mathbf{r} \right) . \tag{5.11}$$

Taking the inner product with $\chi_m^* \left(\mathbf{r} \right)$ in each side,

$$\varepsilon \sum_{l=1}^{M} c_l \cdot \langle \chi_m | \chi_l \rangle = \sum_{l=1}^{M} c_l \cdot \langle \chi_m | \hat{L} | \chi_l \rangle . \tag{5.12}$$

Defining the matrix elements,

$$a_{ml} = \langle \chi_m | \hat{L} | \chi_l \rangle , \tag{5.13}$$

and considering the property of the orthogonal basis set, $\langle \chi_m | \chi_l \rangle = \delta_{ml}$, we get

$$\varepsilon \cdot c_m = \sum_{l=1}^{M} a_{ml} \cdot c_l , \quad i.e., \quad \begin{pmatrix} c_1 \\ \vdots \\ c_M \end{pmatrix} = \begin{pmatrix} a_{11} & a_{12} & \cdots & a_{1M} \\ a_{21} & & \ddots & \vdots \\ \vdots & & & \\ a_{M1} & \cdots & & a_{MM} \end{pmatrix} \cdot \begin{pmatrix} c_1 \\ \vdots \\ c_M \end{pmatrix} . \tag{5.14}$$

Then the rewritten matrix form can be handled by any linear algebra package.[17]

5.4.2   Basis Set Choice Based on Physical Perspective

Considering **what kind of basis sets to be chosen**, it is a mathematical point of view to choose one for which the matrix elements in Eq. (5.13) are easily evaluated. The traditional way has taken, however, more intuitive guidelines looking at Eq. (5.9) that which quantity can be natural to expand the wave function as a superposition from physical/chemical perspective. For periodic solid systems, the natural choice is to expand it by **planewave basis sets** to describe the waves spreading in the solid, while for isolated molecular systems, the wave function would properly expanded as a superposition of atomic orbital functions. In the latter case, each atomic orbital is expressed as a superposition of multiple Gaussian-type functions so that the matrix elements of Eq. (5.13) can be evaluated easily even with the old-time computational techniques, leading to the framework of **Gaussian basis sets** [30].

---

[17] The world's supercomputers are ranked based on the performance of the linear algebra package such as LINPACK. Beginners might have such question, "Why such specific performance to solve simultaneous equations or computing inverse matrices can be the measure of the general-purpose machine? Who wants to solve such 'student's exercise problems' by using supercomputers?" However, this surely makes sense when you realize that most problems in physics/engineering to be solved by supercomputers are usually written in terms of differential equations, which are conventionally handled in terms of basis set expansions.

In Sect. 4.2.2, we mentioned that the handling of the Gaussian basis set is more difficult for beginners than that of the planewave basis set. The past motivation for the Gaussian basis set was the use of "molecular integral tables" [31],[18] but such a requirement has disappeared long times ago by the rapid development of digital computers. Nevertheless, the Gaussian basis has still been popularly used in the molecular science with some rational reason. During the history of the Gaussian basis sets, such knowledge has been accumulated that "for this kind of bonding dealt with the this level of basis sets, we can trust the accuracy up to what number of digits", so it has got difficult to switch to different basis sets easily, especially in the field where **the chemical accuracy**[19] is required.[20]

5.4.3 Basis Sets in Modern Context

The Gaussian basis set assumes to describe such molecular orbitals that have a cusp at the origin of the nucleus.[21] The reason why the cusp was expressed as a superposition of many Gaussian functions was just for the technical reason to utilize "the molecular integral table" mentioned above. With modern computational power, however, it gets feasible to treat directly such atomic orbitals with a cusp (**Slater orbitals**) as a basis set from the beginning to evaluate the matrix elements, Eq. (5.13). Such a "package with Slater basis" (ADF) [32] is actually available.

In the planewave basis calculations, a large E_{cut} is required to describe drastic oscillations of the wave function around the nucleus. Since physical properties we want to describe are regarding to the valence region with smooth spatial variation, it is quite inconvenient to waste the computational cost and capacity for the high E_{cut} that is actually not relevant to the bonding nature. It is therefore a natural course of action to adopt the **mixed basis** where the nuclear part is described by atomic orbitals while only the valence region is treated by planewaves (e.g., APW method [22,33]).

---

[18] By expanding a wave function in a series of Gaussian functions, the matrix elements to be evaluated get reduced to the integrals of each Gaussian product. There is the convenient property that the product of Gaussian functions is a Gaussian function again. The integral evaluation then reduced to follow the propagation relation of each exponent appearing in the product, which is summarized in a book of numerical tables, known as "Kotani table" [31] distributed worldwide.

[19] To distinguish between single and double bondings, it is said that a resolution of the simulations requires about 1 kcal/mol, otherwise any predictions get to be meaningless. The resolution is called the chemical accuracy.

[20] A similar story is found in the use of FORTRAN for numerical simulations. For the field where calibration of numerical accuracy is important, the reliability of the compiler is critical. The accumulation of knowledge on the reliability has been built up on FORTRAN compilers, so it was not easy to switch to other programming languages.

[21] Molecular orbital methods dealing with lighter elements usually adopt all-electron calculations with bare Coulomb potential $1/r$ rather than the pseudopotential. The divergence at $r = 0$ correspondingly produces the cusp at the origin in the orbital function.

The basis sets explained above are introduced based on the physical picture, but, in recent years, there has been a trend toward choosing the optimal basis set from the viewpoint of **numerical efficiency**. Isolated molecular systems have traditionally been treated by the Gaussian basis sets, but recently there has been an increase in the use of the planewave basis set, in which a molecule system is located in a large box which forms a periodic reputation. The idea is that if the box size gets sufficiently large, the intermolecular interaction between neighboring boxes approaches zero, being reduced to an isolated molecular system. By this way, we can avoid the complexity of Gaussian basis handling, but it is a very extravagant usage from the viewpoint of "basis set economics". Reminding that the Fourier transform of $\delta(x)$ is a constant function [19], you can understand that if you try describing a localized/isolated event by the planewaves, it is required to take into account quite wider range of wavelength components.

When something is slightly updated within a local domain, it is a "small thing" in real space intuition, but it is a big issue in the planewave basis set description because the weights over all wavelength components will be affected and required to be updated. In large supercomputers, the speedup is achieved by parallel processing, where the whole calculation is divided with respect to the wavelength components. Then "the update over all the component" leads to the **all-to-all communication** across all the parallel processes.

The all-to-all communication is a greatest cause of performance degradation in massively parallel processing [34]. To avoid this, we want to adopt such basis sets being capable to describe a local event in real space just by the locally updated basis sets (**localized basis sets**). From this viewpoint, the Gaussian basis has desirable property as a representative localized basis set. The spline functions [34] are the expanding basis set which is designed from this viewpoint,[22] namely, the local update of the functional value can be represented by the smaller numbers basis sets as possible. Such a package of the electronic structure calculation using spline basis set expansion is available (e.g., CONQUEST [35]).

The DFT implementation using real space grid (e.g., RSDFT [36]) can be understood as the ultimate in the above context of avoiding all-to-all communication. The implementation is actually replacing differential operations appearing in the equation with the numerical differences on the grid instead of performing basis set expansions as given in Eq. (5.9). In the view of the basis set expansion, however, the real space grid can be regarded as "the expansion using $\delta(x)$", so it can be viewed as adopting the ultimate localized basis.

---

[22] When drawing freehand curves in PowerPoint or other software, you can fine-tune the shape by moving the adjustment points indicated by symbols with the mouse, but you will find that moving one part of the curve smoothly connects to another without much impact. This is where the spline function is used.

5.4.4 Cost Reduction via Basis Set

Denoting the size of the system (number of electrons, orbitals, basis sets, etc.) as N, the planewave basis requires to update all the components ($\sim N$) even for a local event, being disadvantageous in the view of updating cost. Localized basis set is then understood to save the updating cost by $\sim 1/N$ compared to the planewave basis calculation.

Tea Break

In the author's field (QMC electronic state calculations), the spline basis set gets to be popularly used because of its localized nature. Though the orbital functions are originally generated in planewave basis set, their shapes are re-expanded by the spline, otherwise the QMC with planewave is too heavy because of the whole-range update as explained. Though the spline re-expansion brings considerable speedup (around several hundred times faster), it requires much larger memory capacity to store the expanding basis set. The recent technical trend to many-core processors has limited "the memory capacity per core", being disadvantageous for the spline basis set. This stimulates the new direction for implementing the mixed basis set (atomic basis plus planewave) which can be compact in capacity and less-updating nature for a local event.

It is useful to capture how the computational cost increases as the system size increases, scaling to the power of N. The CCSD(T) method is known as the "golden standard" as an accurate method in molecular science, but unfortunately it gets too costly for larger systems as $\sim N^7$ [30]. DFT with planewave basis expansion is said to have a scaling of $\sim N^3$. By using the localized basis set instead of the planewave, the cost reduces by $\sim 1/N$ realizing $\sim N^2$ scaling. The efforts toward further to achieve the scaling down to $\sim N$ are forming the concept of **order-N computation** [22], for the future capability to perform ab initio treatments of the whole device geometry.

Tea Break

While in the precise experimental science, the difference even between the smaller factors matters, there is the concept of "order estimate" mainly in theoretical field where "300 and 700 are roughly the same (because they have the same order of magnitude), while 300 and 3,000 are qualitatively different (because they have different orders of magnitude)". There, N^3 and N^7 are regarded as "qualitatively very different". In the field "computational complexity" (a subject in computer science), however, N^3 and N^7 are recognized as "not much different" (both in the polynomial time/computable). The qualitative difference is only between the polynomial time N^k and $\sim \exp N$ (exponential time/incomputable). It is an interesting contrast depending on the viewpoint of topics to be discussed. Note that even though that seems a matter of "factor difference", "above or below the liquid nitrogen temperature or the melting temperature of the silicon substrate" matters much as a big qualitative difference. For this topic, I remind that a famous experimentalist complained against the order estimate culture of theoreticians claiming "person's heights with 160 cm or 200 cm is surely the trivial matter of the factor, but if one finds a person with 300 cm, it's never the matter of factor".

5.5   Pseudopotentials

The concept of pseudopotential has been introduced in Sects. 3.2.4 and 4.4. Here, we explain the minimum knowledge on the specifications of the pseudopotential required for practical users.

5.5.1   Core/Valence Partitioning

The pseudopotentials describe the effective field felt by the valence electrons, as explained. For such a treatment, how to divide between the core and valence electrons is basically specified by a user, though the partitioning is usually standardized. A user has to choose proper pseudopotentials corresponding to the partitioning as one prefers. For example, taking a case from transition metal elements, the partitioning can be either "1s, 2s, 2p, 3s (core)/3p, 3d, 4s (valence)" or "1s, 2s, 2p (core)/3s, 3p, 3d, 4s (valence)". Namely, there is an optional choice whether 3s electrons are treated as core or valence electrons that should be determined by a user.

Correspondingly, there are pseudopotentials with different numbers of valence electrons even for the same element. It is important to check whether the pseudopotential you have adopted is consistent with your **intended core/valence partitioning**. The pseudopotential with a smaller number of core electrons is called "**small core**", and the one with a larger is called "**large core**". As the 3s in the above example, the orbital which can either be the core or the valence optionally is called "**semi-core**". From the inside, the order is "core/semi-core/valence", where semi-core can optionally be either core or valence.

In such cases with controversy as to whether the semi-core contribution to the bonding is negligible or not, the prediction might be observed to depend on the core size of the pseudopotential, such as "when the semi-core is put in valence, the bond length increases", etc. In one case experienced by the author [37], the choice of exchange–correlation potential was affected by the core/valence partitioning.

Small-core pseudopotentials require more number of electrons to be simulated, and thus increases the computational cost. The ultimate limit is the **all-electron calculations**, in which all electrons are simulated without any pseudopotentials. All-electron calculations are naturally more expensive, but the beginners might have the impression that this is "a sacrifice to be made for accuracy", namely, since core electrons (originally, quantum mechanical particles) are approximated to a fixed potential, the all-electron calculations without such an approximation would be more correct. However, from the viewpoint of experts in numerical calculations, it is difficult to say that the all-electron calculation is accurate.

Core electrons and valence electrons have quite different magnitudes (several orders difference) in their length scale and energy scale. As a common sense of the numerical calculation, it is very dangerous to handle the quantities with very different magnitudes in a simulation simultaneously, getting unreliable results due to the loss of significance [34]. In this sense, the use of pseudopotentials has positive implications to avoid such error making the prediction more reliable from viewpoint of the numerical calculation. In such a most simplified explanation for pseudopotentials, it is explained as "a technique to reduce computational cost by excluding core electrons and dealing only with valence electrons", "by taking only the valence electrons as the simulation target, we can avoid worrying about the increase of the cost as we go down the periodic table (increase of atomic numbers and hence the total number of electrons). It is true, but such a way of the explanation alone would give a risk for beginners to have a reluctant impression for the use of pseudopotentials. Again, the use of pseudopotentials has the positive implications to get reliable numerical results.

The minimum attitude required for users regarding the pseudopotential is to be able to answer to the questions, "how is the core/valence partitioning in your calculation? Namely, how many valence electrons for each element?". Beginners sometimes tend to answer "I don't know the details of the theory", but this is never a matter of methodological experts, but rather the matter in the domain where a user should be responsible for the physical view of the target system.

5.5.2 Practical Categorizing Pseudopotentials

The earlier concept to design the pseudopotentials is to construct the equivalent scattering potential for valence electrons. The earliest implementation along this policy is called **norm-conserving** pseudopotentials (NC/Norm Conserving). Though we omit the detailed explanations here [22], the equivalent scattering can be described by the same amount of the norm (the charge density integrated over within a certain core radius) to be conserved between the original (all-electron) and the pseudodescriptions.

DFT calculations with norm-conserving pseudopotentials have expanded the extent of their application successfully, but they were often faced with cases where the E_{cut} (energy cutoff) requirement described in Sect. 4.2.2 became too large and hence the calculations could not be performed due to memory resource constraints.[23] The larger E_{cut} means that the number of expanding terms in Eq. (4.1) becomes large, requiring more memory capacities to store it, leading to increasing computational cost scaling to a power of the number of expanding terms.

---

[23] The author is acutely aware of this because the quantum Monte Carlo electronic state calculations (the author's majoring method) can only deal with norm-conserving type, being different from DFT.

Remembering the physical meaning of E_{cut} as explained in Sect. 4.2, the larger E_{cut} means that the shorter wavelength component is required to describe the systems with spatially drastic change in the potentials (**hard potential**). To save the computational cost, it is hence necessary to make the pseudopotential be with milder spatial changes (**soft potential**). The hardness of the norm-conserving pseudopotentials is, however, coming from the constraint of the norm conservation. The question is therefore whether it is possible to construct a pseudopotential that avoids the constraint. Though the technical details are omitted here [22], such pseudopotentials are constructed without the norm-conserving constraint because of the above context, being called the **ultrasoft pseudopotentials** (US/Ultra-Soft).

The ultrasoft pseudopotentials can be viewed as an evolving generation over the norm-conserving ones to avoid increasing E_{cut}. There is a further newer generation, the **PAW pseudopotentials** (Projector Augmented Wave), getting practical popularity recently [22]. The details of the background are complicated, but a quick understanding required for practitioners would be "a pseudopotential framework based on the most modern formulation, which can be viewed as an all-electronic calculation".

On a practical level, the ultrasoft can be understood as "a way to reduce E_{cut}", and PAW as "a way requiring no explicit awareness of valence/core partitioning" as possible solutions. Compared to the norm-conserving ones, however, these newer generations of pseudopotentials leave your feeling a bit unclear about its physical picture. This is related to the fact that the initial concept with a concrete physical picture gets more abstract as the generation is evolved in such a way just to achieve required functionalities by any of mathematical framework even without physical picture. To understand this, remember that a similar structure of the evolution was found in Sect. 5.4.3, which discusses modern basis set expansion. The basis sets were originally developed based on physical pictures to get the planewave basis set or the Gaussian basis set. But by focusing on the functionality realized to achieve computational efficiency, they are evolved into the spline basis sets or the real space grid where any specific physical picture is abandoned. When you try to learn the background theory of PAW, it starts with the statement like "Let us consider the linear operator $\hat{T}$ that maps from the all-electron calculation to the valence electron region", that is quite abstract without any physical picture like "the effective scatterer felt by valence electrons" appeared in the norm-conserving context. The concept of pseudopotentials is updated like a "mathematical black box" that reproduces the results of the all-electron calculation equivalently with high computational efficiency even without any physical picture. To realize this, a "mapping linear operator" is set up, and the parameters of it are adjusted to correctly reproduce the all-electron calculation. The inverse operation of the linear operator can also be performed, so that the physical quantities in the all-electron calculation can be reconstructed from the results of the PAW pseudopotential calculation. Once again, it is important to be aware of the conceptual shift where just focusing on the functionality to be realized, the implementation is constructed in an abstract and general mathematical model that does not rely on a specific physical picture. With such awareness, you would be able to go beyond the initial statement with $\hat{T}$ above avoiding frustration.

In the above explanation, "NC"→ 'US' → "PAW" is described as if it was a sort of generational evolution. This might give misleading impression that PAW is always recommended as the latest technique to be applied, but it is not true. As explained in Sect. 1.3.3, very careful approach is required when you start a research on a new substance, not to calculate the target system from the beginning, but starting with a similar system first for which some preceding reports are available for reference. Confirming the agreement of the results with previous reference values, the computational conditions including the geometry are gradually modified toward the final target system with caution. If the previous study used NC, we first perform similar calculations using NC, and then switch to US or PAW after confirming the agreement within NC. In this sense, it is never like "the older generation of pseudopotentials is no longer needed".[24],[25]

5.5.3 Two More Practical Specs of Pseudopotentials

In addition to the three broad categories of NC/US/PAW, other specifications that the practitioners should know at the user level (so that one can answer possible questions at conferences, etc.) include the handling of **reference state** and **relativistic effect**.

For reference state, suppose a pseudopotential for a Ni atom, for example. It is easy to imagine that the Ni site in Ni-metal and that in Ni-oxide have different ionic valence. Then we would expect different "scatterers for valence electrons" in each system, and hence it is reasonable to think that the pseudopotentials of Ni-core for each system are different, describing, e.g., Ni^+ or Ni^{2+}, respectively. Thus, there are various choices available for the pseudopotential, depending on "**which valence state was assumed**". Even for practitioners, it is important to make a user understand what valence state is assumed to be represented by the pseudopotential in one's calculation. As such, it is necessary to be prepared for such questions asking "what is the reference state in your pseudopotentials?" Recently, the conventional way of "assuming and fixing some specific reference state" has evolved further. In the recent implementation of the "**OTF** (on the fly) pseudopotentials", the reference state can be dynamically updated during the calculation according to the charge around the ionic cores [14]. Such an implementation is now available in some DFT packages, getting popular.

---

[24] The similar reasoning also appears in the case of Dirac–Fock or Hartree–Fock pseudopotentials. While the former takes into account the relativistic effects, the latter does not (non-relativistic). Once I've asked to my colleague when I was a post-doc why the Hartree–Fock pseudopotentials are still provided for what purpose because it seems obsolete. The reason was that there are many previous studies using Hartree–Fock pseudopotentials as precise reference values for atoms and molecules, and then it is still necessary for calibrating calculations to check the agreement with these data.

[25] Another reason is that some methodologies only allow the use of NC due to theoretical principles, like QMC electronic state calculations as a typical example. For the QMC, "DFT using NC" is necessary to generate the initial trial wave function for the consistency.

Since relativistic effects are more pronounced in the vicinity of the ion core, where electron kinetic energy is higher, the prescription to consider them effectively at the level of pseudopotentials is reasonable to some extent. To create a pseudopotential, one first calculates the isolated ionic system precisely using an all-electron calculation, and then constructs the pseudopotential to reproduce the equivalent properties (orbital shape, energy levels, etc.) only by the valence electrons and pseudopotentials. Here, there arises a variation of "**which method is used to calculate the isolated system**". For such calculations, DFT is also used in many cases, but in some serious cases, more accurate quantum chemical methods (multi-determinant-based methods) are also used to create pseudopotentials. In such cases, the Dirac–Fock (DF) method is sometimes used to take relativistic effects into account, and the pseudopotentials thus generated incorporate relativistic effects to a certain extent [38]. As another way to incorporate the effects, there are pseudopotentials with LS coupling parameters working as a model-like manner [12].

When using pseudopotentials generated by DFT, the file names of the pseudopotentials are sometimes accompanied with the name of the exchange–correlation functional used to generate it. This often leads to the confusing misunderstanding for beginners that the exchange–correlation functional is a concept relating to the pseudopotential, but as we have explained, it's not true. The exchange–correlation functional and the pseudopotentials are basically independent concepts. In principle, there is no problem to use "pseudopotentials generated by LDA" for GGA calculations. Similarly, "pseudopotentials generated by HF", "ones by DF", "ones by CCSD(T)", etc. may be used in "GGA-DFT calculations".

5.6   How to Choose Appropriate Package for Your Project

5.6.1   Starting with Basis Sets

Beginners are likely to start with the question, "Which package should I use?". Through the explanation about the basis set, one may have noticed that variations of the basis set roughly categorize the software packages. As such, the proper choice should be started with considering whether the system you are working is mainly periodic or isolated systems. Unless any particular reasons exist, "planewave basis" or "Gaussian basis" is the safer choice. For some problems, pseudopotential calculations may not be the appropriate choice (e.g., inner-shell excitations), in which case you come to the choice with the all-electron package at this stage.

Even within the planewave basis set, there still exists several more options. "Is it free", "you have to buy it", "free but you have to register" is also an important factor in the choice. While some, including Quantum Espresso used in this book, are free, others may cost more than ten thousand dollars for a single package. When using popular methods such as planewave or Gaussian basis sets, all packages have almost equivalent functionality, but there are some newer methodologies those are available "only in this package". Then, you can narrow down the choice based on the "exchange–correlation functional/pseudopotential/ basis sets" that you want to use in your project, though these functionalities are expected to be included in any package if you can wait long enough. In the author's experience, such functionalities including MCSCF [30], GW [2,28], and PAW are the ones only available in some packages, but not in others (in 2010s).

5.6.2 Tips for the Choice

In practical aspect, using the "platform with more users" is more beneficial. Just like a familiar example of historical rise and fall in the variety of Internet browsers, it is sometimes the case that the number of users or the market gets poorer to make it obsolete. It is especially critical for beginning practitioners who put more importance on "how to use", ending up with "I learned this, but now it's gone". As the saying goes, "Numbers are strength", and the more users there are, the more strongly they are differentiated from others in terms of richer manuals and user forums.

Tea Break

There was a time in Japanese Academic when there was a strong current to recommend making one's simulation codes to be developed for general-purpose packaging and public release. Correspondingly, there were many domestic packages released and distributed in Japan. However, it takes a lot of energy to maintain manuals and version updates for an unspecified number of users. There are a few packages that are now in a state of "wondering what that person is doing now".

It is difficult for beginners to identify whether a simulation program is truly general purpose or the one just for a temporary use to compose research papers. Even if the true intention is for the latter, they sometimes pretend to take the form of "generalized code open to general users" for getting higher evaluation in competitive research funds. Putting the planewave and the Gaussian basis set methods positioned as "ordinary" methods, "methodologically cutting-edge" methods such as the real space grid, Slater orbital basis set, etc., are sometimes still at the stage where the code is just shared among colleagues to write papers as a research platform for methodological development. They are difficult to be said as the tools for general applications of materials science with non-expert users in any specific methodology.

Tea Break

Once, a student who wanted to join the author's group said "I have downloaded a DFT package, and now started to learn how to use it". I asked which package he was working on, and he replied the name of package which was unique in its methodology developed in Japan. I instructed him that the one was actually not regarded as a general-purpose one, and made him switch to other conventional package. This kind of incident sometimes occurs because beginners tend to take the ones which have the manuals written in Japanese, being possible to be accessed by Google search in Japanese.

5.6.3   Kernel, Package, and Wrapper

Other than Quantum Espresso, the following packages are representative ones used as general-purpose packages with a larger number of users internationally[26] :

- VASP (charged) [16]: Planewave basis set DFT.
- CASTEP [14]: Planewave basis set DFT. For academic use, the **kernel** source code is available for free. Otherwise, one have to purchase its binary accommodated in the package "Materials Studio" [13].
- SIESTA (for free) [39]: Localized basis set DFT.
- ABINIT (for free) [40]: Planewave basis set DFT.
- CRYSTAL (charged) [41]: Gaussian basis set DFT for periodic systems.
- GAUSSIAN (charged) [42]: Gaussian basis set DFT and MO (molecular orbital methods).
- GAMESS (for free) [43]: Gaussian basis set DFT and MO (molecular orbital methods).
- NWCHEM (for free) [44], PYSCF (for free) [45]: Gaussian basis set DFT and MO (molecular orbital methods).
- ...

As in the above example of "Materials Studio" $\supset$ "CASTEP", the kernel name (CASTEP) may have been widely known than the package name (Materials Studio) due to its historical background. Unless one is an expert in the methodological research field, one might be confused by these names to understand what is the relation between the kernel and the package. For our package treated in this book, the relation is "Quantum Espresso" $\supset$ "PWSCF", which kernel name "PWSCF" is now not used much anymore.

In more updated usages, we have to mention the relationship between **wrappers**[27] and packages to avoid misunderstandings. There are several utilities to evaluate specific physical quantities that calls ab initio packages inside of their script structures (as explained in Fig. 1.2). Well-known example is "Phonopy" [15] to evaluate phonon properties. By writing the relationship as "Phonopy" $\supset$ "[kernel name]", it can switch between VASP and Quantum Espresso as the [kernel]. Similarly, there are "LOBSTER" [46] (for bonding analysis), "USPEX" [48] (for structure search using genetic algorithms), etc. as wrappers that call "VASP" [16] or "Quantum Espresso" [6] as its kernel upon the user's choice.

---

[26] There are also some Japanese packages that are steadily increasing their users with long-lasting maintenance.

[27] The upper framework that "wrap's up the lower kernels for each functionalities.

References

1. "Fundamentals of Condensed Matter Physics", Marvin L. Cohen, Steven G. Louie, Cambridge University Press (2016/5/26) ISBN-13:978-0521513319
2. "Electron Correlation in the Solid State", N.H. March (ed.), Imperial College Press (1999), ISBN:1-86094-200-8
3. "Quantum Monte Carlo simulations of solids" W.M.C. Foulkes, L. Mitas, R. J. Needs, and G. Rajagopal Rev. Mod. Phys. 73, 33 (2001). https://doi.org/10.1103/RevModPhys.73.33
4. "*Ab initio* search of polymer crystals with high thermal conductivity", K. Utimula, T. Ichibha, R. Maezono, K. Hongo, Chem. Mater. 13, 4649-4656 (2019). https://doi.org/10.1021/acs.chemmater.9b00020
5. "Potential high-Tc superconductivity in $YCeH_{20}$ and $LaCeH_{20}$ under pressure", P. Song, H. Zhufeng, K. Nakano, K. Hongo, R. Maezono, Mater. Today Phys. 28, 100873 (2022). https://doi.org/10.1016/j.mtphys.2022.100873
6. https://www.quantum-espresso.org (URL confirmed on 2022.11)
7. "First-principles anharmonic vibrational study of the structure of calcium silicate under lower mantle conditions", J.C.A. Prentice, R. Maezono, and R.J. Needs, Phys. Rev. B 99, 064101 (2019). https://doi.org/10.1103/PhysRevB.99.064101
8. "Synergy of Binary Substitution for Improving Cycle Performance in LiNiO2 Revealed by ab initio Materials Informatics", T. Yoshida, R. Maezono, and K. Hongo, ACS Omega 5, 13403-13408 (2020). https://doi.org/10.1021/acsomega.0c01649
9. "Candidate structure for the H2-PRE phase of solid hydrogen", T. Ichibha, Y. Zhang, K. Hongo, R. Maezono, F.A. Reboredo, Phys. Rev. B 104, 214111 (2021). https://doi.org/10.1103/PhysRevB.104.214111
10. "Computational Physics", Jos Thijssen, Cambridge University Press (2007/3/22) ISBN-13:978-0521833462
11. "Bandgap reduction of photocatalytic TiO_2 nanotube by Cu doping", S.K. Gharaei, M. Abbasnejad, R. Maezono, Sci. Rep. 8, 14192 (2018).
12. "Quantum Monte Carlo study of atomic and solid sodium", R. Maezono, M.D. Towler, Y. Lee and R.J. Needs, Phys. Rev. B 68, 165103:1-9 (2003). http://dx.doi.org/10.1103/PhysRevB.68.165103
13. https://en.wikipedia.org/wiki/Materials_Studio (URL confirmed on 2022.11)
14. https://en.wikipedia.org/wiki/CASTEP (URL confirmed on 2022.11)
15. https://phonopy.github.io/phonopy (URL confirmed on 2022.11)
16. https://www.vasp.at (URL confirmed on 2022.11)
17. "Exploring Chemistry With Electronic Structure Methods: A Guide to Using Gaussian", James B. Foresman, Gaussian (1996/8/1) ISBN-13:978-0963676931
18. "Density-functional Theory of Atoms And Molecules" (International Series of Monographs on Chemistry), Robert G. Parr, Yang Weitao, Oxford University Press USA (1989/1/1), ISBN-13:978-0195092769
19. Ken F. Riley, Mike P. Hobson, Stephen J. Bence, "Mathematical Methods for Physics and Engineering" (Third Edition), Cambridge University Press,
20. "The Story of Spin", University of Chicago Press (1998/10/1), ISBN-13:978-0226807942
21. "Correlation and Localization", Peter R. Surjan, Springer (1999/8/15), ISBN-13:978-3540657545
22. "Electronic Structure: Basic Theory and Practical Methods", Richard M. Martin, Cambridge University Press (2004/4/8), ISBN-13:978-0521782852
23. "Density Functional Theory: A Practical Introduction", David S. Sholl, Janice A. Steckel, Wiley-Interscience (2009/4/13), ISBN-13:978-0470373170.
24. "The Theory of Intermolecular Forces", Anthony Stone, Oxford Univ Press (1997/12/4), ISBN-13:978-0198558835
25. "Anomalous non-additive dispersion interactions in systems of three one-dimensional wires", A.J. Misquitta, R. Maezono, N.D. Drummond, A.J. Stone, and R.J. Needs., Phys. Rev. B 89, 045140:1-9 (2014). http://dx.doi.org/10.1103/PhysRevB.89.045140

26. "A computational scheme to evaluate Hamaker constants of molecules with practical size and anisotropy", K. Hongo and R. Maezono, J. Chem. Theory Comput. 13, 5217-5230 (2017). https://doi.org/10.1021/acs.jctc.6b01159
27. "Lecture Notes on Electron Correlation and Magnetism", Patrik Fazekas, World Scientific (1999/5/1) ISBN-13:978-9810224745
28. "Strong Coulomb Correlations in Electronic Structure Calculations", Vladimir I Anisimov (ed.), CRC Press (2000/5/1) ISBN-13:978-9056991319
29. "Classical Electrodynamics", John David Jackson, Wiley (1998/8/14), ISBN-13:978-0471309321
30. "Molecular Electronic-Structure Theory", Trygve Helgaker, Poul Jorgensen, Jeppe Olsen, Wiley (2013/2/18), ISBN-13:978-1118531471
31. "Table of Molecular Integrals", M. Kotani, A. Amemiya, E. Ishiguro and T. Kimura, Maruzen, Co., Tokyo (1955) ISBN-13:9780598976918
32. https://www.scm.com/doc/ADF/Input/Basis_sets_and_atomic_fragments.html (URL confirmed on 2022.11)
33. "Solid State Physics", N.W. Ashcroft and N.D. Mermin, Thomson Learning (1976).
34. "Computational Physics: Problem Solving with Python", Rubin H. Landau, Manuel J. Paez, Cristian C. Bordeianu Wiley-VCH (2015/9/8) ISBN-13:978-3527413157
35. http://www.order-n.org (URL confirmed on 2022.11)
36. https://www.rsdft.jp (URL confirmed on 2022.11)
37. "GaN band-gap bias caused by semi-core treatment in pseudopotentials analyzed by the diffusion Monte Carlo method", Y Nikaido, T. Ichibha, K. Nakano, K. Hongo, and R. Maezono, AIP Adv. 11, 025225 (2021). https://doi.org/10.1063/5.0035047
38. "Smooth relativistic Hartree-Fock pseudopotentials for H to Ba and Lu to Hg", J.R. Trail and R.J. Needs, J. Chem. Phys. 122, 174109 (2005)
39. https://departments.icmab.es/leem/siesta/ (URL confirmed on 2022.11)
40. https://www.abinit.org (URL confirmed on 2022.11)
41. https://www.crystal.unito.it/index.php (URL confirmed on 2022.11)
42. https://gaussian.com (URL confirmed on 2022.11)
43. https://www.msg.chem.iastate.edu/gamess/ (URL confirmed on 2022.11)
44. https://www.nwchem-sw.org (URL confirmed on 2022.11)
45. http://pyscf.org (URL confirmed on 2022.11)
46. http://www.cohp.de (URL confirmed on 2022.11)
47. "Optimization of Many-body Wave function" R. Maezono, J. Comput. Theor. Nanosci., 6, 2474 (2009). https://doi.org/10.1166/jctn.2009.1308
48. https://uspex-team.org/en (URL confirmed on 2022.11)

Part III
Advanced Topics

This part outlines various directions on how the topic can be expanded upon the understanding of the kernel part explained so far (Chap. 6). In the spirit of keeping the content as minimal as possible, we have limited the subject matter to bulk solids in the preceding chapters. Most of practitioners are, however, interested in how to model and capture practical problems such as impurities and surfaces as real-world problems. The part also explains such modeling. It also explains how the energy and charge density as direct output of the kernel calculation can be used to evaluate practical properties with respect to the response of the materials.

As mentioned in Chap. 1, the application of electronic structure calculation has been greatly advanced by devising the outer loop that invokes the kernel calculation. By applying data scientific methods to the outer loop, electronic structure calculations have evolved into 'Materials Informatics'. The concept of representative methods for this part (genetic algorithm/particle swarm algorithm/Bayesian search) is outlined in Chap. 7.

Chapter 8 discusses tips for successful management of collaborative projects, taking into account the peculiar nature of *ab initio* analysis.

Section 11.5 is derived from Chap. 6. Although it explains an advanced topic, they are useful for a deeper understanding of the band theory.

Toward Practical Applications

6

Abstract

So far, we have explained why we have to learn first how to evaluate the fundamental properties (DOS and band dispersion) of pristine bulk, using the metaphor of bread dough (Sect. 3.1), or in the context of the practical procedure starting from trace calculations (Sect. 8.1.4). However, most of the readers are likely to be motivated to pick up this book by a specific mission to deal with practical topics such as "magnetism of molecules placed on a solid surface" or "ionic diffusions of impurities in a bulk", etc. Starting from the evaluation of fundamental properties explained so far, we are then interested in how is the connection toward the practical missions in the aspects of

- (a) toward more advanced geometries (surfaces, defects, etc.),
- (b) toward more advanced properties (response, reactions, transition temperatures, etc.).

In this chapter, we will explain these connections. In the latter part of the chapter, we will learn about the impact of the acceleration of the kernel calculations and the concepts of "high throughput" and "workflow" in this context.

6.1 Toward Advanced Geometries

6.1.1 Construction of Geometries

Modeling of Surfaces

In the tutorials so far, we prepared the input geometries of bulk structures by downloading the cif-files as explained in Sect. 3.2.1. Unlike the bulk cases, we have to prepare the geometries of surfaces and defects by ourselves as we want by using the **structure modeling tools** such as Materials Studio [1], etc. The tools can provide

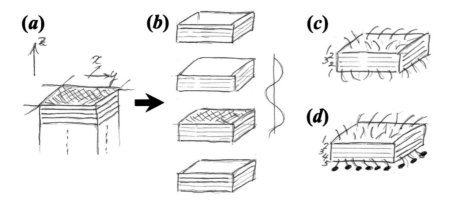

Fig. 6.1 The periodic slab model used to describe the surfaces. Let z-direction be the normal of the surface extending in xy-plane [panel (**a**)]. Atomic layers in the plane are actually repeated along the negative z-direction toward infinity, but it is modeled by the slabs with five layer that are separated by vacuum layers repeatedly [panel (**b**)]. There are two options in the way that how many atomic layers are represented by the five-layered slab. The naive way is to regard the five layers labeled as "12321" in symmetric way [panel (**c**)] to describe three atomic layers with the third as the deepest layer from the surface. The hair-like whiskers growing from the surface schematically represent the bonding hands from the surface atoms. A black circle drawn at the tip of the whiskers represents the terminations of a bonding hand by, e.g., hydrogen atoms, making the surface inactivated for any reactions. The panel (**d**) hence models the five atomic layers from the surface by the five-layered slab, with the layer "5" being the deepest inside atomic layer

the surface models as specified, such as "three-layered model with the (111) surface direction cut out from a mother bulk structure".

The model of surfaces (**periodic slab model**) commonly used in ab initio [2] calculations is explained in Fig. 6.1. Let z-direction be the normal of the surface extending in xy-plane [panel (a)]. Atomic layers in the plane are actually repeated along the negative z-direction toward infinity, but it is modeled by the slabs with five layers that are separated by vacuum layers repeatedly as shown in the panel (b). As the vacuum separation gets larger, the model gets closer to the isolated surface if the number of layers (e.g., five) is large enough. The reason why such a periodic geometry is adopted is because we want to describe the problem using the planewave basis set. This is a commonly used strategy for planewave calculations (e.g., an isolated molecule modeled by a periodic array of "a molecule in a large box".

For the five-layered model in Fig. 6.1, there are two options in the way that how many layers are represented by the "five layers in the slab". The naive way is to regard the five layers labeled as "12321" in symmetric way [panel (c)] since both sides of a slab face to vacuum. In this case, the five-layered slab describes the three atomic layers with the third as the deepest layer from the surface [3]. In panels (c) and (d), the hair-like whiskers growing from the surface are depicted which schematically represent the bonding hands from the surface atoms. In another option described by the panel (d), a black circle is drawn at the tip of the whiskers, which represents the terminations of a bonding hand by hydrogen atom, etc. On the black-circled side of the surface, therefore, the bonding hand is blocked, resulting in an atomic layer that has been inactivated and has completely lost its ability to react [atomic layer "5" in panel (d)]. The inactivated layer can be used as the deepest inside atomic layer, and hence the terminated slab "12345" describes five atomic layer from the surface.

We mentioned the modeling tools which provide the input geometries as specified with the surface orientation and the number of layers required in the slab. Some additional manual work is then performed to place surface terminations and adsorbed molecules located on the slab, which can be done as intuitive as dragging and attaching images of atoms on the GUI. This modified slab is then multiplied by the structural optimization described in Sect. 5.1.2 to create a relaxed structure, which can then be periodically arranged as shown in Fig. 6.1b to represent desired surfaces [1].

Modeling of Defects, Impurities, and Alloys

As a large fraction of the usage of ab initio simulations in Materials Science, the prediction how properties are modified by impurities and defects is an important mission. Geometries with impurities are prepared by making the elemental substitutions on the atomic sites of the pristine bulks. Instead of elemental substitutions, we can remove the fraction of atoms from atomic sites that corresponds to the structural model with defects [4]. For interstitial defects, you can put extra atoms in between the atomic sites by using the structure modeling tools (e.g., Materials Studio).

In these treatments, it is important to note the "**charge neutrality compensation**. If you try substituting sites with ions of different valence, there will be an excess or deficiency of charge. Here the knowledge that users should keep in mind is that the treatment of systems with no charge neutrality in periodic systems is very tricky being an issue in the domain of experts [5]. A periodic system consists of an infinite number of unit cells, so if one unit cell has a non-zero charge, the resultant system to be considered becomes an infinitely charged system, which is mathematically difficult to deal with. As such, when substituting ions of different valence, the calculation is performed by constructing a unit cell that maintains charge neutrality by making holes available elsewhere to compensate for the neutrality condition, or by introducing additional elements with valence that counteract the excess or deficiency [6].

Another concept to keep in mind when introducing "impurities" such as defects and substitutions is the use of **supercell** [4], which bundles several times the number of unit cells for the "no impurity (pristine) system". Suppose that the pristine unit cell with length L [panel (a) in Fig. 6.2] is subject to the replacement of an atomic site by an impurity as "x→o" [panel (b)]. If we naively consider "xxoxx" as a unit cell of the impurity system, the impurity will be repeated with a period of length L, leading to spurious interactions between mirror images of impurities that does not actually exist. In order to treat the impurities by the planewave basis set, the impurities must appear periodically, but in order to eliminate false impurity interactions, the distance between periodically appearing impurities must be separated enough. For this purpose, the unit cell of the impurity system is taken as multiples of the pristine unit cell to form enlarged one (supercell) (in the example in Fig. 6.2, three times larger supercell is used).

Fig. 6.2 To prepare the unit cell with impurities (x→o) based on the pristine unit cell [panel (**a**)] that with the same size spuriously introduces the interactions between mirror images that does not actually exist [panel (**b**)]. To exclude that the cell size should be taken larger so that the mirror images get separated enough [panel (**c**)]

(a)
```
    <-L->
...|xxxxx|xxxxx|xxxxx|...
```

(b)
```
...|xxoxx|xxoxx|xxoxx|...
```

(b)
```
...|xxxxx|xxoxx|xxxxx|xxxxx|xxoxx|xxxxx|...
   |<------ 3L ----->|...                |
```

In Sect. 5.4.4, we mentioned the computational cost of usual planewave basis set DFT being scaling to $\sim N^3$. If we adopt, say, $3 \times 3 \times 3$ supercell for an impurity/defect system,[1] then the computational cost is estimated, getting around $\sim 27^3$ times larger. Users have to be aware that the treatment of atomic site substitutions (doping, defects, solid solutions, etc.) requires much increased computational cost due to the enlarged unit cell.

! **Attention**

As we mentioned, the use of a supercell for advanced geometry requires a large computational cost (CPU time as well as memory capacity). This situation would be the most common entry for a beginner to consider for using a supercomputer. For DFT calculations, there used be a problem that the parallelization efficiency (see also Sect. 6.3.3) was not very high, but at present, it gets much improved to achieve the acceleration linearly up to ~ 100 processors.

In addition to the increased computational cost due to the enlarged supercell, there is a more serious problem in dealing with the atomic site substitutions. That is the problem of explosion of "number of cases" for the substitution configurations. Even for a fixed concentration of substitutions, the number of possible configurations sometimes reaches hundreds of millions to trillions [4, 7]. Although they can be reduced to a few hundred or a few thousand at most by the space group theory, it is practically impossible to generate all of the trillions of cases written out for the group theory classification. For the affordable number of the whole possibility to be accommodated within the available memory capacities, several structural modeling tools such as Materials Studio provide the ability to perform the group theoretical classification to narrow down the number of structural models. In many practical problems in materials science, however, a number of cases may go easily beyond the memory/file capacity. There is a tool to cope with the difficulty which prevents to write out all the configurations but generates only group-theoretically inequiv-

---

[1] It refers three times larger cell in each direction with respect to the pristine unit cell.

alent structural models based on the modern combinatorics theory by saving the memory/file capacities [4].

6.1.2   Finite Size Error

As explained at the end of Sect. 4.2.1, the $M \times M \times M$ k-mesh in a calculation corresponds to $M \times M \times M$ times the unit cell adopted as the simulation cell in the real space. If the unit cell of the structure model itself gets larger (e.g., already K times of the pristine unit cell for doped systems), it is computationally difficult to perform the calculation using further K times multiple of the unit cell to form simulation cells to see the convergence with respect to the k-mesh as we did in Fig. 4.2. In this case, we have no choice but to use, the result of an affordable calculation, say, up to $2 \times 2 \times 2$ at most, which includes the bias from the result of $M \times M \times M$ (M is a large enough integer). The bias due to the insufficient size to capture the finite extension is called **finite size error** [8].

For expensive calculations where the k-mesh convergence cannot be evaluated, the estimations are taken as the predictions by affordable smaller simulation cells plus the corrections for finite size error. Various schemes for the corrections of finite size errors have been developed for this purpose [8,9]. One source of finite size error is the bias about the kinetic energy due to the fact that the coarse k-grid cannot capture the longer wavelength components as explained in Sect. 4.2.1. There is another source of bias caused by description of the interaction. Smaller simulation cells introduce the spurious interactions between the adjacent mirror images repeatedly appearing in the array of simulation cells, leading to false contribution depending on the size of the simulation cell but vanishes when the simulation cell gets larger. A similar story was mentioned in Fig. 6.2 as spurious interaction between impurities, but even if the cell does not include any impurity, the electron itself interacts with mirror images of itself in the neighboring cell. Developing such scheme to eliminate the bias is itself an advantaged research topic in the quantum many-body problem [8,9]. A topic which is accessible as an extension of the contents of this book is the correction of the finite size bias by the twist averaging as given in the appendix Sect. 11.5.

6.2   Toward Evaluations of Advanced Properties

6.2.1   Sorting Out the Form of Computational Evaluations

Practical properties that the readers are originally interested in, such as "spectrum/conduction/catalytic reactivity/ etc.", seem to be tremendously far apart from what we learned so far, i.e., the fundamental properties such as "ground state energy/dispersion diagram/density of states". How these are connected? The calculations of physical properties using ab initio calculations can be broadly classified into the following four categories:

- **(1)** Quantities justified fundamentally to be evaluated by DFT.
- **(2)** Quantities not so well justified fundamentally but DFT evaluations are optimistically expected to be working.
- **(3)** Quantities evaluated by the formulae in model theories with the parameters provided by DFT calculations.
- **(4)** Quantities predicted based on the data correlation with the quantities that DFT can evaluate.

In the previous chapter, the readers have learned that the DFT implemented in practical packages is originally justifying the energy and charge density in the ground state.[2] Kohn–Sham orbitals and their levels are introduced just for mathematical convenience, not strictly for physical significance. Since the band dispersion diagram is just a plot of the eigenvalues of the Kohn–Sham orbitals, one could say, "What physical meaning is there in that? it is just a plot of intermediate quantities in the calculation process...". That's true from the most stoic standpoint that only "the integral value over the occupied Kohn–Sham level is a quantity that has meaning as a ground state energy. "Can we discuss any physics from the unoccupied levels that appear there?" is the naive question from the most stoic viewpoint.

In practical use, however, the band dispersion diagram obtained by DFT is used to discuss, e.g., "the absorption of light at the transition from this occupied band to this unoccupied band", etc. As mentioned in Sect. 5.3.4, DFT with such an exchange–correlation functional, "0% correlation and 100% exchange" roughly corresponds to the Hartree–Fock theory as an extreme case.[3] In Hartree–Fock theory, one-body orbitals and their energy levels are the quantities that can be physically meaningful. Considering this connection to the Hartree–Fock theory at the extreme point of exchange–correlation selection, we can optimistically think that the dispersion diagram of the Kohn–Sham levels can also have physical meaning to some extent, allowing us to make such discussions like "the response due the excitation from this occupied level to this unoccupied level", etc. Such an optimistic and aggressive application of the band diagram to discuss excitation, response, etc., refer "(2)" in the above category.

> **! Attention**

It is now well known that the DFT tends to underestimate the bandgap [2, 11]. In some cases, the gap is closed rather than underestimation. When the author was a graduate student, this "gap problem" was too negatively pointed out as something like "fundamental break down of DFT". Of course, the methodological community of DFT has known that "it does not make much sense to make a fuss about the Kohn–Sham gap compared to the experimental gap", but when DFT gets to be known

---

[2] There are such studies to develop further extension of DFT beyond the ground state as a fundamental challenge [10].

[3] This is actually "the exact exchange DFT", not the Hartree–Fock theory in exactly speaking, but quite similar to each other.

for non-experts of methodology, there was a surprise that the calculated dispersion diagram was so close to the experimental ones for semiconductors. This led to the misunderstanding that the Kohn–Sham gap should be close to experiments, which is a sort of tragedy.

The quantities categorized as (1) above are, therefore, the physical properties that can be evaluated from the ground state energy values, while those categorized as (2) are the quantities evaluated from the values of Kohn–Sham levels especially including those of unoccupied levels. For instance, estimating elastic constants and phonon frequencies from the change in the ground state energy caused by slightly applying distortion to the geometry belongs to the category (1).[4] Also, determining the crystal structure to minimize the ground state energy and obtaining the XRD pattern (X-ray diffraction spectrum) as its Fourier transform [12,13] also belongs to the category (1).

Taking the optimistic standing point (2) to put the physical meaning for the unoccupied states of the Kohn–Sham level as excited levels, then we are feasible to evaluate spectra and response coefficients. Since these quantities are in the form of the golden rule type,

$$\chi(\omega) = \sum_{b,\mathbf{k}} A_{b',\mathbf{k}';b,\mathbf{k}} \cdot \delta\left[\left(\varepsilon_{b',\mathbf{k}'}^{(u)} - \varepsilon_{b,\mathbf{k}}^{(o)}\right) - \omega\right], \qquad (6.1)$$

they can be evaluated in the form of "fishfinder", i.e., counting up the possibilities distributed among a band dispersion diagram: Eq. (6.1) can be read as the counting up over the non-zero contributions of each δ function with the weight A. The non-zero contributions comes where the energy difference between the unoccupied excited level (superscript <u>underline</u>unoccupied) and the occupied level (<u>o</u>ccupied), $\left(\varepsilon_{b',\mathbf{k}'}^{(u)} - \varepsilon_{b,\mathbf{k}}^{(o)}\right)$, matches the specified excitation energy ω". $\chi(\omega)$ therefore measures how "dense" the possible targets ("fish") are distributed over all modes $(b, \mathbf{k})$ for a given ω [11].

The category (3) [formulae in model theories with parameters evaluated by ab initio calculations] may sound somewhat anticlimactic, as it did to the author when I was a beginner. When I heard that "thermal conductivity can be calculated by an ab initio package, I wondered what kind of framework was being used in the "ab initio evaluation of the conductivity". Actually, I found that it is simply a matter of calculating and substituting phonon frequencies and the coefficients of the higher order anharmonic terms by fitting the dependence of ab initio energy depending on the deformations. These numerical values are required to evaluate the formula for the conductivity which itself is derived from Boltzmann theory and the related model theories using diagrams based on the linear response theory [19]. The ab initio

---

[4] Note again that we are just talking about the justification in terms of the theoretical framework, exempting it from approximation errors regarding the selection of the exchange–correlation functional.

package has a utility to substitute the ab initio evaluations of the parameters into the formula to get the conductivity (Fig. 1.2). The situation is similar for the evaluation of T_c in BCS superconductors [20]. The category (2) can also be viewed as a special type of the category (3) since it is the model theory that leads to the golden rule-type formula.

The category (4) [Predictions based on correlations] is based on model theories, phenomenological theories, and empirical rules in materials science. For example, when considering work such as estimating the melting point of a solid using ab initio calculations, one wonders how such a dynamical property can be handled by a theory that can only handle the ground state. It is known from phenomenological theory that thermodynamic quantities diverge at phase transition points such as the melting point [14], so the melting point can roughly be estimated by plotting the temperature dependence of thermodynamic quantities calculated by substituting phonon frequencies estimated from ab initio calculations into the formula from model theory and then identify the temperature where the dependence is likely to diverge as "the estimation of melting point". Although numerically accurate values cannot be expected, this approach is effective enough for the purpose of grasping trends such as "whether the melting point gets increased or decreased when doping or composition is modified. This kind of approach will be discussed further in Sect. 6.2.2.

6.2.2   Prediction Based on Data Correlation

Ab initio calculations can predict various physical properties by giving a fixed structure with considerable computational cost. If we regard the property Y evaluated for a given structure X represented as the relation on the XY-plane, we naturally come to the idea to construct a regression relationship by interpolating the selected points obtained by the costly ab initio calculations. By using the regression, we can then predict the property Y for any X with preventing from the costly calculation. One of the major missions of "Materials Informatics" can be said to provide such **regression relation**. In order to improve the accuracy of prediction based on interpolation, one revises the regression by adding data points. To achieve the best efficiency to revise it, one would wonder which point should be augmented by another ab initio calculation. To answer this, Bayesian prediction and other statistical theories [15] can be applied (Sect. 7.2.3).

By the way, although we mentioned "candidate structure X and physical property Y" in a very conceptual manner, when we think about it carefully, we immediately realize that "how to characterize them" is itself a rather complex problem. For structure X, complex crystal or molecular structures cannot be described as simply as "uniaxial X". Considering X to determine a materials property $Y(X)$, it would include various information such as interatomic distances, angles, compositions, etc. being not a "uniaxial X". Of course, multivariable treatments in the theory of regression of $Y(X)$ are well established in Statistics, but, first of all, we have to identify which quantities to be chosen as relevant explanatory variables X (**descriptors**). Such research to perform to select descriptors is an important topic in Materials

Informatics (Sect. 7.3.3). To reduce the number of relevant descriptors for X, such methods using penalty terms (in the context of "sparse modeling") [16], dimension reduction technique such as the auto-encoder [13], etc. have been developed.

Not only for the descriptors X, but also for the **objective variable** Y that it is never be expressed in a very single axis. Strictly within the scope of ab initio DFT, feasible Y to be evaluated would include the bandgap, the quantities based on the ground state energies (formation energies, energy barriers, etc.) However, the recent trend in Materials Informatics has been demanding more useful properties as Y, such as catalytic capacity, plastic deformation strength, and melting point. If one has completed to read the previous chapters, one would be well aware that the only quantities that ab initio DFT can reliably evaluate are those for the ground state. The demanded Y is then completely out of our capability in principle. If we only do what we can do with certainty within this range, then the quantities that ab initio calculations can handle as objective variables Y will be severely limited, and we will end up with a tool that is of no practical use at all.

Here, it is important not to consider Y as the feasible evaluations only within ab initio but rather to consider it as the combination with model theories and various phenomenological theories developed in Materials Science, as described in Sect. 6.2.1. For example, the topic of crack suppression of the solder that holds the device was mentioned in Sect. 1.2.1. In that topic, the capability of the suppression (Y) is finally reduced to the evaluation of the energy barrier for the ionic migration (feasible to be evaluated by DFT), assisted by the combination with phenomenological theories in regarding the cracking and with the theory of migration kinetics. Here, it is important to realize that the absolute value itself is not necessary to be reproduced precisely, but it is enough to know the "trend" (increase or decrease caused by the modification?).[5]

Let us denote "$Y \sim Z_1$" to describe the relation where the trend of the property Y is mainly dominated by the property Z_1. If we can connect $Y \sim Z_1 \sim Z_2 \cdots \sim Z_n$, and Z_n can be calculated by ab initio calculation with a structure accompanied by descriptors X, then we can construct the regression between Y and X by combining $Y \sim Z_n$ and $Z_n(X)$ (**correlation modeling**). For example, let us consider the properties related to dielectric polarization such as optical response (Y). For the magnitude of the polarization (Z_1), there is a model theory (Lyddane–Sachs–Teller relation [17]) stating that the magnitude is related to the "LO-TO splitting of phonon dispersion (Z_2)". As such, we would be able to estimate the trend how the optical response changes qualitatively by modifying the composition of a material via ab initio phonon calculation (Sect. 7.4.2) monitoring the LO-TO splitting in the obtained dispersion, giving $Y \sim Z_2$ [18].

Thermal conductivity [19] or superconducting transition temperature [20] is evaluated by the ab initio calculations of parameters substituted into formulae given by

---

[5] This viewpoint conversion is a kind of "easy to understand once it is pointed out, but difficult to notice by oneself". Since ab initio simulations were originally developed with an awareness of quantitativeness, it has been difficult for methodology developers in particular to realize it, they tend to pursue the quantitative coincidence. The conversion of the viewpoint becomes popular when practitioners start to use DFT as a tool for Materials developments.

model theories (Sect. 1.2.3). These would be regarded as an extreme case of the correlation modeling, $Y = f(Z_n)$ (Z_n being the phonon frequency), where the relation "$\sim$" is replaced by "$=$" and $f(\ldots)$ is the model formula.

Unlike the case with "$Y = f(Z_n)$", the relationship "$Y \sim Z_n$" is not usually based on any firmly established relationship. Rather than finding it in an established textbook, the relation usually comes from the cutting-edge discussions by practitioners. The easiest way to find such relationship would be making Google search putting the words, e.g., "DFT photocatalyst", and imitate the approach discussed in preceding papers. One can find several ab initio works dealing with, e.g., "catalytic activity of photocatalytic solids", but there are so many different approaches even with a fixed target problem and a fixed method. Some work deals with it by analyzing the anisotropy of the catalytic activity from the anisotropy of the energy barrier for ionic migration [21], while others by evaluating the "band lineup" (comparisons of valence/conduction band edge over several compounds taking the vacuum level as common reference level) compared with the water-splitting level [22]. In the latter, one has to consider how to estimate the vacuum levels and the chemical potentials. You might wonder if there would be some firm theoretical framework to evaluate them. But when you read preceding papers dealing with that, you may get such impression that "Are we really sure to adopt such a quite rough and intuitive estimation?". The framework used around here would still be unestablished, and it is expected to grow in the future as a field that can be developed in cooperation with computational thermodynamics [23].

6.3   What the Speed of Simulation Brings

As we have mentioned in Sect. 1.4.2, recent ab initio calculations have become fast enough for single-point calculations, and the script constructions with outer loops providing input files with changing parameters are getting feasible to perform the evaluations of various property, such as phonon calculations and structure exploration. In this context, we mention here the related concepts toward **high throughput** and that of **workflow**.

6.3.1   High-Throughput Handling

Scripts can be used to automate the process of automatically generating input files at high speed, feeding them to kernel calculations, and automatically processing the results. Nowadays, users do not need to build such scripts themselves, and script sets are distributed as a package for common processes such as phonon, dielectric calculations, etc.

Regarding a "call" of such a package as a "one-point calculation of the property",[6] repeating the "one-point calculation of the property" under different conditions would provide a regression relationship described in Sect. 7, leading to a form of Materials Informatics. However, in order to construct a regression relationship with statistically meaningful reliability, a large number of "one-point calculation of a property" must be performed to get enough plotting points. The key to success is how fast and efficient one can perform such a large number of calls (**high throughput**) in order to build such knowledge.

The "single-point **kernel** calculation" is accelerated by **parallel processing over processor resources** for a single job. On the other hand, to speed up the "single-point calculation of a **property**", upper level of parallel processing over **job processing resources** for many jobs should be taken into consideration.[7] Unlike the resource within a CPU processor, "job processing resources" such as "how many computers one owns", "how powerful supercomputer one can access", "what kind of job-classes available on your supercomputer" vary from user to user. The key for more efficiency hence depends on the skill to compose scripts as explained below to handle job managements on each individual available computational resources [24].

The required scripts for efficient job management include that automatically submits jobs to available resources one after another, that automatically recognize which job class is free while which class is full/crowded, that automatically monitors which jobs are completed while which ones still running, that automatically saves the minimum set of I/O into backup storage, etc. To compose such scripts, one should integrate required commands (login/file transfer/job-submission/ staging/backup/etc.) into a script file in order. In the context of Materials Informatics, it is required to automate to decide what to calculate next based on the present pool of existing results to augment the knowledge. For this purpose such scripts that automatically extracts required information from output files from completed jobs and performs required calculations (e.g., calculating energy differences from the energies extracted from output files) to decide the next point to be calculated.

As more updated trend on handling computational resources for high throughput, general-purpose frameworks[8] are getting popular, and one might be required to become proficient in using these in future.

Tea Break

It is often to be asked by the students who are considering to join our group that "What kind of programming language do I need to learn?" Rather than the "programming" language typically like "C++", the "script" language like "Python", etc. are required

---

[6] Recall that a "one-point calculation of a property" is obtained by repeating many "one-point kernel calculations" (Fig. 1.2).

[7] For this level of "parallelization", imagine how to distribute the divided loads for a single project over ten working colleagues, among which some person can process the load faster while some works slowly. Optimizing the way of distribution makes the project achieved faster.

[8] As representative ones, "AiiDA" [25] and "fireworks" [26] are getting popular now.

for students to be proficient. Unless one was involved as a core member to develop the kernel code project like "Quantum Espresso", it is rare nowadays to work on the kernel source code with "programming" language, that would be the case for the readers of this book (practitioners in Materials Science as application users).

6.3.2 Workflow Automation

For such cases with T_c evaluations [20] or thermal conductivity evaluations [19], the whole evaluation requires to call several kinds of utilities (Sect. 1.5.2) for ground state energies, phonon frequencies, anharmonic vibration evaluations, etc. in some certain order to integrate the results. Such a **procedure over applications and utilities** forms a **workflow** that is desired to be automated by the script framework, hopefully.

When established at the earliest stage, such a procedure is regarded as a research originality in its own right. Once the procedure gets to be commonly used, then it should be made into a workflow and provided in a form that can be used by anyone, hopefully. By this way to make it a module, we can concentrate on upper level integrations to deal with more practical properties.

Tea Break

The present readers of this book would be likely to be aware of the keyword "Python". Python is a scripting language, but it gets to be the most widely used mainly because of the richer usability of their libraries especially in the AI-related field. For example, if one wants to work on Neural network, one does not have to build one's own codes, but is recommended to get a set of libraries written in Python downloaded from the Web. Almost all the functions that are frequently used in the field are available in the library, and one can call the function to perform it. "The re-utilization of the accumulated knowledge constructed by others" is strongly recommended in the contemporary developments in Science and Technology, and it just so happens on the Python platform, making it the most popular.

6.3.3 Quantum Monte Carlo Electronic Structure Calculations

Quantum Monte Carlo electronic structure calculations would be the most appropriate example to explain the concrete image of the workflow concept. This method, also called ab initio quantum Monte Carlo, is an approach to achieve the "most reliable electronic structure calculation" by evaluating many-body wave functions handled by Monte Carlo sampling technique [27].

The most critical point in the electronic structure calculations is how to handle quantum many-body interactions as explained in Sect. 5.3.1. In DFT, it takes the strategy of "mapping into an effective one-body form", where the quantum many-body interactions is expressed in model-wise by the exchange–correlation potentials. It leads to the prediction bias depending on the choice exchange–correlation potentials taken from the available variations (Sect. 4.5). DFT has expanded beyond the fields where it was initially applied to realistic materials, i.e., covalent solids, going toward the systems with intermolecular interactions, surfaces, impurities, magnetism, etc. Then the problem of the prediction bias due to the choice of exchange–correlation potentials gets more prominent to be pointed out [28]. These motivate the basic research to further improve the descriptiveness of the exchange–correlation potentials within the framework of DFT.[9]

As other approaches to work on many-body wave functions under interactions, there are two representative frameworks not relying on exchange–correlation potentials, one of which is the molecular orbital (MO) method [29] and the other is the quantum Monte Carlo (QMC) electronic structure calculation [27]. The former (MO) aims at a more precise description of many-body wave functions based on one-body orbitals in more sophisticated way to describe the many-body form, while the latter (QMC) takes the strategy with **sampling** of the many-body wave functions rather than **describe** it.[10] The QMC is inherently a computationally expensive approach, and has been intractable for the problems with practical system size until the massive parallel computation becomes popular. Since the method has extremely high parallelization efficiency (e.g., nearly 1,000 times faster when 1,000 parallel cores are used), it practicality gets rapidly increased, being one of the most reliable ab initio methods applicable to the problems with practical system size [27].

Tea Break

DFT gets more descriptiveness by improving the exchange–correlation functional. It is metaphorically likened to the biblical description of "ascending Jacob's ladder" (Fig. 6.4). The QMC, on the other hand, improves its reliability by performing random walks, i.e., drunken walks. It is interesting that there is a Japanese way to describe the drunken people doing "bar-hopping" as "ascending the ladder".

Compared to DFT, the QMC electronic structure calculations require more complicated procedure [27, 30] as explained below (Fig. 6.5). First, we prepare the orbital functions used to construct anti-symmetrized product (Sect. 5.3.1) obtained by either DFT or MO method (01/Generating orbitals). We further prepare a variational trial function multiplied by a factor called the Jastrow function (02/Construction of the trial function). The Jastrow function contains tunable variational parameters, which works to fine-tune the amplitude of the wave function depending on the electron configuration. This set of parameters is numerically optimized to lower the energy using

---

[9] Remember that the framework of DFT itself is rigorous to ensure the exact description of the normal ground states.

[10] **The curse of dimensionality** [31] implies that it is better to take the strategy of "sampling" rather than "trying to describe" a multidimensional space such as a many-body wave function.

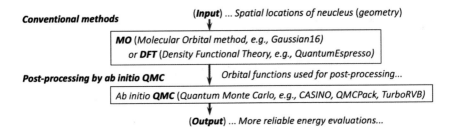

Fig. 6.3 In practical applications, quantum Monte Carlo electronic structure calculations are often used in the form of a post-process that is applied at the end of previous methods such as molecular orbital and density functional methods, and this is where the workflow process of "calling and processing various types of kernel calculations" comes into play

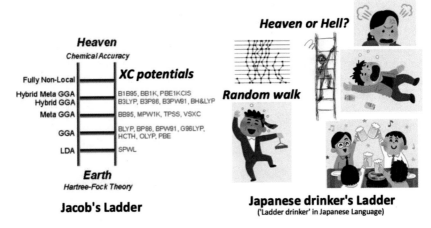

Fig. 6.4 DFT gets more descriptiveness by improving the exchange–correlation functional, being metaphorically described as "ascending Jacob's ladder". The QMC, on the other hand, improves its reliability by performing random walks, i.e., drunken walks. There is a Japanese way to describe the drunken people doing "bar-hopping" as "ascending the ladder" (parts of the figure are taken from [27,33])

the variational Monte Carlo (VMC) method (03/Variational Optimization/VMC). Once a certain degree of optimization is achieved, the optimized parameters are used as the initial trial function. The initial function is further driven by the imaginary time evolution to get deformed toward the exact solution (04/Imaginary Time Evolution/DMC). Each of the steps from 01 to 04 is further subdivided into smaller steps. To perform these steps for the publication level, it currently requires the skill of a "master craftsman", and this surely hinders DMC from getting more popular. To make the method more spread over many users, the procedure should be automated to some extent.

QMC electronic state calculations have a remarkably high parallelization efficiency, making it the most powerful method to demonstrate the performance of massively parallel supercomputers. Although it is still difficult to perform practically more demanding calculations (structural optimization, evaluations for practical

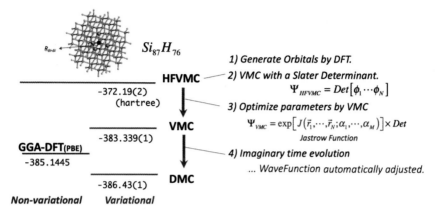

Fig. 6.5 Detailed procedure [(1)∼(4)] inside of the QMC electronic structure calculation which is expressed as a single black box in the previous Fig. 6.3. In QMC method, the energy values (shown in the figure) can be used as a measure how close to the exact answer (the lower, the closer), called "variational principle" [30]. Note that GGA value (DFT in general) is not the case of "variational principle" (the lower does not mean the closer to exact solution)

quantities, etc.) that DFT can, QMC surely offers the highest predictive reliability without such uncertainty depending on the choice of exchange–correlation functionals. As explained in the next chapter, even rather *boring* quantities that are composed of ground state energy (formation energy, etc.) are still important enough to serve as a descriptor in the context of Materials Informatics. As such, the QMC is expected as a promising tool to provide such fundamental quantities for materials database at the highest reliability with less ambiguity assisted by the massive power of national flagship parallel supercomputers. As a "killer app" for such national flagship machine, QMC electronic structure calculation is considered to be applied to all combinations of elements comprehensively to get fundamental quantities in automated manner [32]. This is the background why the workflow of the QMC has been developed to be automated.

6.3.4 What the Simulation Speed Brings

There would be some "critical value" for the degree of simulation acceleration, which qualitatively changes the nature of the research area. For example, if we talk about the acceleration achieving "six times faster", it may

- (a) 12 months gets to 2 months,
- (b) 60 min gets to 10 min,
- (c) 3 s gets to 0.5 s,

but each has different impact for us. The "critical value" is the one comparable to the time scale for human thought that acceptable to keep one's motivation. In this sense, the case of (a) and (c) would end up with "huh!, that's nice", but the case (b) is a big one. Since there will be many other interrupting tasks in a day, a single calculation taking more than half a day would be difficult to make practitioners maintain their motivation for seeking the optimal condition, except those with quite tough ability of motivation management. On the other hand, if the single calculation takes only a few minutes, much more practitioners can easily get into the enthusiasm by changing the parameters and trying next in a trial-and-error manner, saying "Well, what if I change this part?"

The author has worked on the QMC electronic structure calculations using flagship-level supercomputers, where a calculation used to require sometimes 6 months to get a single value of the ground state energy. Hearing that the QMC provides the most reliable predictions for the problems where conventional DFT is not working well, a number of interested researchers approached for collaborations with the author. However, collaboration is unlikely to be successful on such a time scale taking half a year, during which the collaborators' motivation got stalled and they shift their interest to other problems. Even if we obtained the results 6 months later, the collaborators may have forgotten that they were interested in the result when they approached to us in the first place.

However, if the result can be given within a day, then the collaborative discussion can proceed at a high tempo, talking like "Under these conditions, we predict this result", leading to successful collaborations. Once again, we can see how important it is to get such accelerated simulations achieved within **the acceptable time scale** comparable to that maintainable human's motivation. DFT also had a time when "it took several hours to calculate a single point even on a supercomputer". It is after the method becomes feasible to be performed on a laptop computer within a few minutes that the number of users get exploded, spreading over the applications of practical materials. The explosion is not stopped at the level of "just an acceleration", but leading to the qualitative change of the research field combined with Information Science into Materials Informatics as explained in the next chapter.

References

1. https://en.wikipedia.org/wiki/Materials_Studio (URL confirmed on 2022.11)
2. "Density Functional Theory: A Practical Introduction", David S. Sholl, Janice A. Steckel, Wiley-Interscience (2009/4/13), ISBN-13:978-0470373170.
3. "Adhesion of Electrodes on Diamond (111) Surface: A DFT Study", T. Ichibha, K. Hongo, I. Motochi, N.W. Makau, G.O. Amolo, R. Maezono, Diam. Relat. Mater. 81, 168 (2018). https://doi.org/10.1016/j.diamond.2017.12.008
4. "Application of canonical augmentation to the atomic substitution problem", G.I. Prayogo, A. Tirelli, K. Utimula, K. Hongo, R. Maezono, and K. Nakano, J. Chem. Inf. Model 62, 2909-2915 (2022). https://doi.org/10.1021/acs.jcim.2c00389
5. "Electronic Structure: Basic Theory and Practical Methods", Richard M. Martin, Cambridge University Press (2004/4/8), ISBN-13:978-0521782852
6. "Stabilization mechanism of the tetragonal structure in a hydrothermally synthesized $BaTiO_3$ nanocrystal", K. Hongo, S. Kurata, A. Jomphoak, M. Inada, K. Hayashi, R. Maezono, Inorg. Chem. 57, 5413 (2018). https://doi.org/10.1021/acs.inorgchem.8b00381
7. "Stochastic estimations of the total number of classes for a clustering having extremely large samples to be included in the clustering engine", K. Utimula, G.I. Prayogo, K. Nakano, K.

Hongo, and R. Maezono, Adv. Theory Simul. 4, 2000301 (2021). https://doi.org/10.1002/adts.202000301

8. "Finite-size errors in continuum quantum Monte Carlo calculations", N.D. Drummond, R.J. Needs, A. Sorouri, and W.M.C. Foulkes Phys. Rev. B 78, 125106 (2008). https://doi.org/10.1103/PhysRevB.78.125106

9. "Diamond to beta-tin phase transition in Si within quantum Monte Carlo", R. Maezono, N.D. Drummond, A. Ma, and R.J. Needs, Phys. Rev. B 82, 184108 (2010). https://doi.org/10.1103/PhysRevB.82.184108

10. F. Perrot and M.W.C. Dharma-wardana, Phys. Rev. B 62, 16536 (2000).

11. "Fundamentals of Condensed Matter Physics", Marvin L. Cohen, Steven G. Louie, Cambridge University Press (2016/5/26) ISBN-13:978-0521513319

12. "Machine learning clustering technique applied to powder X-ray diffraction patterns to distinguish compositions of ThMn$_{12}$-type alloys", K. Utimula, R. Hunkao, M. Yano, H. Kimoto, K. Hongo, S. Kawaguchi, S. Suwanna, R. Maezono, Adv. Theory Simul. 3, 2000039:1-9 (2020). https://doi.org/10.1002/adts.202000039

13. "Feature space of XRD patterns constructed by auto-encorder", K. Utimula, M. Yano, H. Kimoto, K. Hongo, K. Nakano and R. Maezono In press, Adv. Theory Simul. (2022). https://arxiv.org/abs/2005.11660

14. "Thermodynamics and an Introduction to Thermostatistics (English Edition) Herbert B. Callen, John Wiley & Sons (1985/8/29) ISBN-13:978-0471862567

15. "Synergy of Binary Substitution for Improving Cycle Performance in LiNiO$_2$ Revealed by *ab initio* Materials Informatics", T. Yoshida, R. Maezono, and K. Hongo, ACS Omega 5, 13403-13408 (2020). https://doi.org/10.1021/acsomega.0c01649

16. "First-Principles Study of Structural Transition in LiNiO$_2$ and High Throughput Screening for Long Life Battery", T. Yoshida, K. Hongo, and R. Maezono, J. Phys. Chem. C 123, 14126 (2019). https://doi.org/10.1021/acs.jpcc.8b12556

17. "Solid State Physics", Giuseppe Grosso, Giuseppe Pastori Parravicini, Academic Press (2000/4/15) ISBN-13:978-0123850300

18. "Structural, electronic, and dynamical properties of Pca21-TiO$_2$ by first principles", M. Abbasnejad, M.R. Mohammadizadeh and R. Maezono, Europhys. Lett. 97, 56003:1-6 (2012). https://doi.org/10.1209/0295-5075/97/56003

19. "Ab initio search of polymer crystals with high thermal conductivity", K. Utimula, T. Ichibha, R. Maezono, K. Hongo, Chem. Mater. 13, 4649-4656 (2019). https://doi.org/10.1021/acs.chemmater.9b00020

20. "Potential high-Tc superconductivity in YCeH$_{20}$ and LaCeH$_{20}$ under pressure", P. Song, H. Zhufeng, K. Nakano, K. Hongo, R. Maezono, Mater. Today Phys. 28, 100873 (2022). https://doi.org/10.1016/j.mtphys.2022.100873

21. "Ti interstitial flows giving rutile TiO$_2$ reoxidation process enhanced in (001) surface", T. Ichibha, A. Benali, K. Hongo, and R. Maezono, Phys. Rev. Mater. 3, 125801 (2019). https://doi.org/10.1103/PhysRevMaterials.3.125801

22. "Bandgap reduction of photocatalytic TiO$_2$ nanotube by Cu doping", S.K. Gharaei, M. Abbasnejad, R. Maezono, Sci. Rep. 8, 14192 (2018). https://doi.org/10.1038/s41598-018-32130-w

23. "*Ab initio* thermodynamic properties of certain compounds in Nd-Fe-B system", A.T. Hanindriyo, S. Sridar, K.C. Hari Kumar, K. Hongo, R. Maezono, Comp. Mater. Sci. 180, 109696 (2020). https://doi.org/10.1016/j.commatsci.2020.109696

24. "Jisaku PC cluster tyou-nyuumon", Ryo Maezono, Morikita Publishing (2017/12/14) ISBN-13:978-4627818217

25. https://www.aiida.net (URL confirmed on 2022.11)

26. https://materialsproject.github.io/fireworks/ (URL confirmed on 2022.11)

27. "Quantum Monte Carlo simulations of solids" W.M.C. Foulkes, L. Mitas, R. J. Needs, and G. Rajagopal Rev. Mod. Phys. 73, 33 (2001). https://doi.org/10.1103/RevModPhys.73.33

28. "The Importance of Electron Correlation on Stacking Interaction of Adenine-Thymine Base-Pair Step in B-DNA: A Quantum Monte Carlo Study", K. Hongo N.T. Cuong, and R. Maezono, J. Chem. Theory Comput. 9, 1081 (2013). https://doi.org/10.1021/ct301065f

29. "Molecular Electronic-Structure Theory", Trygve Helgaker, Poul Jorgensen, Jeppe Olsen, Wiley (2013/2/18), ISBN-13:978-1118531471

30. "Size dependence of the bulk modulus of semiconductors nanocrystals", R. Cherian, C. Gerard, P. Mahadevan, N.T. Cuong and R. Maezono, Phys. Rev. B 82, 235321:1-7 (2010). https://doi.org/10.1103/PhysRevB.82.235321

31. "The Elements of Statistical Learning: Data Mining, Inference, and Prediction", Trevor Hastie, Robert Tibshirani, Jerome Friedman, Springer (2009/3/1) ISBN-13:978-0387848570

32. https://www.mgi.gov (URL confirmed on 2022.11)

33. "General Performance of Density Functionals", S.F. Sousa, P.A. Fernandes, and M.J. Ramos, J. Phys. Chem. A 2007, 111, 42, 10439 (2007). https://doi.org/10.1021/jp0734474

Materials Informatics Based on Structural Search

7

Abstract

We mentioned structure search and geometry optimization briefly in Sect. 5.1.1. Now that the single-point calculation gets to be fast enough, which leads to the concept of **the structure search** by by randomly and rapidly generating structural input files using outlying scripts to find a structure achieving lower energy heuristically. This type of simulation is found to be powerful for predicting the crystal structure under extreme conditions (e.g., under ultra-high pressure such as under the crustal mantle) where experiments are difficult to perform. Such studies have reported newly predicted structural composition that has a great influence on the theory of geophysics. With this impact, the ab initio structure search has led to one flow of useful applications of ab initio calculation. Following the approach with randomly updating the structure as the first generation, more sophisticated strategies have been developed to improve the sampling efficiency by utilizing the 'knowledge' over previous samples that achieved good scores. In this chapter, we will discuss the following representative strategies for the improvement:

- Particle swarm optimization,
- Genetic algorithm,
- Bayesian inference,

It is one of the major flow which links the ab initio calculations with the field of data science to form a new field, Materials Informatics.

© The Author(s), under exclusive license to Springer Nature Singapore Pte Ltd. 2023   173
R. Maezono, *Ab initio Calculation Tutorial*,
https://doi.org/10.1007/978-981-99-0919-3_7

7.1   Formulation of Materials Search

7.1.1   Descriptors to Introduce Search-Space

In[1] **the materials structure search**, each structure should be represented as a single point 'X' in a search-space. The mission is then to find the optimal X to achieve better value in the evaluation function $Y(X)$,

$$[X; \text{Descriptor space}] \rightarrow [Y; \text{Evaluation function}] . \tag{7.1}$$

When representing a material structure as 'X', we are first faced with the problem of how to represent it the single point X in a multidimensional space. For example, in the case of a crystal structure, we need to specify it by a set of information such as crystal system, independent pairs of lattice constants, coordinates of atomic positions inside the Bravais lattice, atomic numbers of elements contained, etc. If these pairs of information can be quantified numerically instead of verbally, it would be possible to express them as a single point in a multidimensional space. For example, instead of using the word 'orthorhombic crystal', we need to replace it with some kind of integer number as an index.

However, since the number of parameters required to define the structure differs between cubic and orthorhombic crystals, they cannot be treated in the same dimensional space. When searching for the advantage of cubic or orthorhombic crystals, the comparison must be developed on the same footing, so a vector space that simply lists the parameters for each crystalline system cannot handle the problem.

Various researchers have already established achievements by their efforts to represent material structures as multi-parameterized **descriptors**. The products of these efforts have been published as libraries which are available on web as we can use. For example, 'XenonPy' [1] is a utility that generates descriptors for a given chemical composition, composed from 58 properties such as atomic number, atomic radius, etc., of 94 elements as weighted by its composition ratio. It provides us hundreds of descriptors based on the combination of 58 characteristics automatically, but instead of using all of them as they are, we narrow down the descriptive components (trimming of descriptors) and use them in a smaller number of dimensions. We will discuss how to do this in another section.

'XenonPy' is a descriptor for crystal structures, but descriptors for complex molecular structures used in organic chemistry are well developed in the field of **Chemoinformatics**, and a prominent example is the SMILES representation [2]. Molecular structures are described by connecting element symbol alphabet with characteristic bonding structures such as rings, single bonds, double bonds, etc., which are converted into character strings according to a convention and expressed as **SMILES**

---

[1] E.g., if it predicts a mineral containing hydrogen at a composition ratio that has not been known so far being stable under extremely high pressure, the amount of hydrogen that can exist on the earth will change. This gives a great impact on the argument on 'where is the unfound hydrogen on the earth' by providing the possibility to find it in the crust.

representations. The framework to treating word strings in a search space has been well developed in the field of natural language processing. Once the molecular structures are mapped into word strings, then the framework can be used to develop our materials search [3].

The careful reader would notice that both 'XenonPy' and 'SMILES' represent the composition rather than the structure as a point in space. There are cases where the same composition gives different structures just like degenerated information. This will be discussed later (Sect. 7.4.1).

The descriptors given by 'XenonPy' can only represent up to composition, but one can add more components on the descriptor to incorporate crystal system numbers, physical properties, etc. [4]. Though the original context was 'material structure represented as a single point in the search space', such extended descriptors can make it to be developed further toward the concept of 'the search for desired properties in the descriptor space' rather beyond a mere structure search.

7.1.2 For Efficiency to Sample Search-Space

Beyond **the random sampling** [5], which is a fast random generation and feeding of structural input files, we will discuss particle swarm optimization, genetic algorithm, and Bayesian search as representative strategies for higher efficiency. The details of each method are described in separate subsections, but in this section, we will outline them, comparing each other.

As a general problem of spatial search, what is required in methods is how to cope with the situation where multiple peaks of the evaluation function exist around the space apart from each other. A search method (e.g., steepest descend) is a framework to propose the next sampling position so that it directs toward more coziness (lower energy/higher evaluation function). In this case, the question is whether there is a more comfortable location much further away from the current comfortable area (**global optimization**). In the conventional gradient method, the search is performed based only on local information in the form of gradients evaluated at that location, it gets to be caught easily by **local minima** and has a weakness to capture global minima apart from the current location. To overcome this, various tricks have been studied, e.g., 'once a local minimum is found, take a big jump out of it', etc., and these strategies are widely called **meta-heuristics** [6].

In the gradient method, the local gradient is used as the **utilization** of the current knowledge to determine the next step. Random sampling corresponds to the global **search** without the utilization, just by the random choice for the next. When compared to random sampling, each of the methods described in this chapter can be contrasted in terms of 'utilization' and 'search' as follows.

- **Random sampling**

    - (search) Parallel-wise multiple sampling
    - (utilization) N/A

- **Particle swarm**

  - (search) Parallel-wise multiple sampling
  - (utilization) Determining the direction toward the superior point by a directed vector to make the next samplings direct to it.

- **Genetic algorithm**

  - (search) Parallel-wise multiple sampling with the possibility of mutation.
  - (utilization) Generating the next samplings based on the current samplings with superior scores with crossover operation.

- **Bayesian inference**

  - (search) Considering the weight of prioritizing unknown regions in next-step decisions
  - (utilization) Also considering the weight of prioritizing the larger evaluation function.

In particle swarm [7], multicomponent in a descriptors is treated as the component of a directed vectors and 'utilization' is achieved via orientation of the vector. In genetic search [8], multicomponent is regarded as an array rather than a vector. The 'utilization' is realized by multiplying eugenic arrays to determine the next, the 'search' is devised by generating global jumps through tricks such as 'crossover in recombining arrays' or 'mutations' that mimic those in biological genes. In Bayesian search [9], the estimation of the evaluation function is treated in terms of 'mean + error'. This method balances 'utilization' and 'search' by taking a linear combination of two weights, each of which accounts for the trend toward 'higher evaluation function (mean)' and 'higher uncertainty (error)' to determine the next sampling points.

7.1.3   Regression to Describe Properties

Once the correspondence $X \rightarrow Y$ as Eq. (7.1) becomes clear, it leads to the idea of regressing the $Y(X)$ relationship. Up to this point, Y has been the energy value so that we may seek energetically stable unknown structures X as in the random search.

Conceptually, we can upgrade $Y(X)$ from the energy value to those such as 'superconducting transition temperature T_c' or 'deviation from the desired band gap value', etc., so that we can search for structures that realize the desired properties, as the ultimate goal of material design.

Though it is basically possible to use ab initio calculations to evaluate Y fully as in the manner described in Sect. 6.2, such a full evaluation is too costly to perform the structure search. Therefore, the idea is to seriously evaluate only a few points by ab initio calculations and interpolate the rest with regression relations. Or, forget about ab initio calculations for the moment and build the regression by learning from the vast amount of 'structure/property' data pairs in the available database.

Descriptor variables $\mathbf{X}$ are usually taken to be about 50 components in practice [4]. Such a high-dimensional regression has well been developed in the field of information science, such as binary-tree regression and neural network regression as representative frameworks [10]. From the word of 'regression', one might remind of the theory of linear regression as one learned it in undergraduate [11]. Unlike such classically rigid theories, the above modern data regressions are actually such numerical implementations of a black box to reproduce a give relation $Y(\mathbf{X})$ by adjusting a number of parameters inside. Although its difficult name may intimidate you, it is important to take it easy and master its use. Classical regression is a rather elaborate quite mathematical framework to achieve fit with a small number of parameters, since it is a method developed in the time that we cannot take advantage of the computational power. Modern data regression, on the other hand, pushes the difficulty to numerical optimization of black-box parameters, so its conception is not so difficult, as explained in Sect. 7.3.

Details on how to use the constructed regression are given in Sect. 7.4. When one thinks of using regression, one usually thinks of traversing the regression graph backwards to find the domain X that achieves the desired Y. However, the descriptor space X was not necessarily continuous, since it reflected the material structure. If we are considering lattice constants or bond angles within a particular crystalline system, it can be continuous, but in the current context, we are considering global comparison even over different crystal structures or compositions in a single space, so the $\mathbf{X}$ axis would never continuous. In such cases, the inverse mapping $Y \to X$ is not necessarily well-defined.

Tea Break

In the fundamentals of analysis taught in undergraduate math, there was a theorem called 'the intermediate value theorem' [11]. When we completed its proof at length, some of us might be puzzled by the fact that what we had worked so hard to prove was 'graphically obvious'. The gist of the lengthy proof was actually to concern about what would happen when the traced back point on x were not exist in the operation $y \to x$.

The question is how to use the constructed regression to find optimal X that achieves the desired range of properties in a forward search rather than a backward search (as depicted in Fig. 7.1 conceptually). This is where the technique named in Sect. 7.1.2 is applied. If the value tapped out from the regression is used as the evaluation function, it is possible to bring the sampling points to optimal X that maximizes or minimizes the target quantity of property in question.[2]

---

[2] To make it toward a 'desired value', taking the deviation of the regression value from the target value to be minimized in the evaluation function would work.

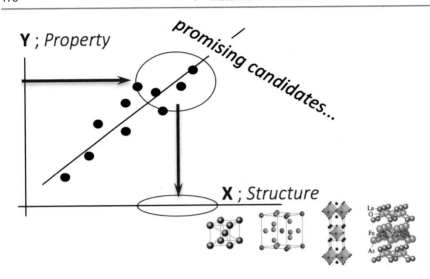

Fig. 7.1 The naive image is to identify the X (structure) that realizes the desired Y (property) by inverse estimation as depicted, but there is not always a corresponding structure existing in the range (e.g., when X is non-continuous like isolated islands). In practice, inverse estimation is achieved by combining forward estimations, not by backward tracing back as depicted

In the construction of a regression relation with Y as the property, we noted that it does not necessarily require ab initio calculations, but existing database for materials properties. Is there a place for the inevitable use of ab initio calculations? Structure prediction methods may predict a composition or structure that is expected to achieve a desired property. However, whether the found optimal structure is indeed stable is not considered in general. Therefore, as a final step, we need to check whether the predicted structure is indeed energetically stable using ab initio calculations (Sect. 7.4).

7.2   Search Methods

7.2.1   Particle Swarm Optimization

The simplest view for multicomponent in descriptors would be to regard them as components of **a directed vector** in a multidimensional vector space. Another view is to regard them as **a multi-component array**, such as genetic information. The former picture is adopted in **the particle swarm algorithm** (PSO) [7] while the latter in **the genetic algorithm**, which will be discussed in the next section.

As outlined in Sect. 7.1.2, the particle swarm algorithm performs a spatial search by scattering a large number of search points throughout the space. It implements the use of 'vested knowledge' in such a way that the entire group of search points flows in the direction of the point where each search point has scored the highest so far. By considering a multi-component descriptor as a multi-component directed

Fig. 7.2 The star shows the past best position $\mathbf{x}_g$ of the whole group and the white square is the past best position $\mathbf{x}_p^{(i)}$ of each individual. The inertia vector $\mathbf{v}^{(i)}[k+1]$ to determine the next search position $\mathbf{x}^{(i)}[k+1]$ is modified from the direction of the previous inertia $\mathbf{v}^{(i)}[k]$ toward the past best

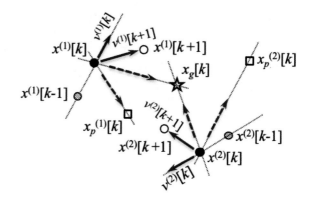

vector, the picture of 'heading in the direction of ...' can be used. The algorithm is said to mimic a flock behavior, such as a flock of birds being led by the behavior of a single bird [7].

Let i be the index of each sampling point, $\mathbf{x}_p^{(i)}$ the past best position of each individual i, and $\mathbf{x}_g$ the past best position of the whole group. The knowledge of $\mathbf{x}_g$ is assumed to be shared by all individuals. Each individual determines its next-step position $\mathbf{x}^{(i)}[k+1]$ reflecting its inertia $\mathbf{v}^{(i)}[k]$ at the current time (k-step). Its inertia direction is to be updated from $\mathbf{v}^{(i)}[k]$ by modifying it as follows (w below is the weight factor).

$$\mathbf{v}^{(i)}[k + 1] = w_v \cdot \mathbf{v}^{(i)}[k] + \mathbf{f}_{\text{bias}} . \tag{7.2}$$

The modifying bias flow $\mathbf{f}_{\text{bias}}$ is given as

$$\mathbf{f}_{\text{bias}} = w_p \cdot \left(\mathbf{x}_p^{(i)}[k] - \mathbf{x}^{(i)}[k]\right) + w_g \cdot \left(\mathbf{x}_g[k] - \mathbf{x}^{(i)}[k]\right) , \tag{7.3}$$

as a balanced linear combination between 'the flow towards each individual best $\mathbf{x}_p^{(i)}$' and 'the flow toward the group best $\mathbf{x}_g$', weighted by $w_{g,p}$ (Fig. 7.2).

By the updated inertia $\mathbf{v}^{(i)}[k + 1]$, the next sampling position is updated as

$$\mathbf{x}^{(i)}[k + 1] = \mathbf{x}^{(i)}[k] + \mathbf{v}^{(i)}[k + 1] , \tag{7.4}$$

as shown in Fig. 7.2. The weights $\{w_v, w_p, w_g\}$ are chosen randomly within the extent of each given amplitude, which is set by a user. There are several Various manners to set the weights have been proposed by researchers [7]. The above PSO strategy to update the searching sampling points has been implemented successfully so that it can be applied to material structure descriptors, now widely used [12].

7.2.2   Genetic Algorithm

As explained at the beginning of the previous subsection, we consider multi-component descriptors as information arrays rather than directed vectors, and take the idea to mimic how biological gene sequences recombine and evolve as an optimizing update of the array [8]. In genetic adaptation to the environment, the 'utilization' of the vested knowledge so far corresponds to the generation of the next genes based on the previous genes with superior scores by the crossover, and the 'search' to overcome multimodality is realized by the mutation of the array. These 'genetic operations' are summarized as

- Selection
- Crossover
- Mutation

as explained below.

There are several manners of operation to leave only high-scoring sequences in **selection** operation, such as 'leaving N sequences from the top (ranked selection)' or 'selecting stochastically with probability according to the score (roulette selection).

In the **crossover operation**, the array sequences of the next generation (children) are created from those of parent generation as follows.

Let the two parents to create a child have the sequences 'abcdefghi' and '$\alpha\beta\gamma\delta\varepsilon\phi\psi\chi\omega$'. For example, the array of the child is created by separating 'abc | defg | hi' at two *crossing points* and cutting and pasting the two parent arrays as 'abc | $\delta\varepsilon\phi\psi$ | hi'. There are various manners for how many crossing points to take depending on each implementation, and the choice of the crossing points is usually randomly taken.

The simplest form of bf the mutation operation is to randomly select some region in the array of a parent and replace it with other components. The *inversion* is a commonly used implementation for the mutation, as explained below. Selecting an area in the array (e.g., 'efg' in the above example) and inverting the array order to create a child ('abcd | efg | hi→'abcd | gfe | hi'). In this example, however, it is necessary to assume that the array components $\{e, f, g\}$ are substitutable for each other. For example, if this $\{e, f, g\}$ is responsible for the 'crystal system/lattice constants/binding angles', they never be substitutable having different dimensions and different definition ranges. Some preprocessing would be necessary to enable inversion for this case.

The above is a brief outline of genetic algorithms. For the operations 'selection/crossover/mutation', there are numerous options on how to implement them, and methodological research on which one is superior for what kind of problem is still ongoing. Often, beginners are intimidated that there might be a serious theoretical framework to be learned such as the framework of Physics or Chemistry behind the application of various genetic manipulations. However, there is no such rigorous and deductive guiding principle in particular, and for the most part, random recombination rules such as the roulette method are applied in a trial-and-error manner for efficiency [8]. Although the name of the method often sounds intimidating, it is a good idea to take it easy and learn as you use the code implementation.

Code implementation of genetic algorithms to predict crystal structures has become much more practical and widely used in recent years [13,14]. Note that its use is slightly different from the picture described in this subsection, for which we provide a separate subsection later in Sect. 7.4.1.

7.2.3 Beysian Search

To realize a global search without any uncovered range, particle swarm search and genetic search are implemented with multiple sampling points. Bayesian search, on the other hand, can basically realize the global search even with a single sampling point, though it can be extended to multiple sampling in a straightforward manner.

In Bayesian search, a numerical measure of unexploredness $\sigma(x)$ is given at each point x to be searched [9]. It adopts the strategy that the next sampling point is decided by the priority with higher unexplored score $\sigma(x)$. It also considers another weight to prioritize the higher value of the evaluation function $\mu(x)$, but if there are areas with low μ but high σ, they will be given priority as well, by setting $A(x) = \mu(x) + \lambda \cdot \sigma(x)$ with an adjustable the weight λ.

For the unexplored score, it is set to zero at sampled points in the past as the known points, and the finite variance σ is assigned centered on the interpolated value of the evaluation function in the region between the known points, as shown in Fig. 7.3. This variance is used as the unexplored score. For the variance σ, the Gaussian distribution is assumed around interpolated values. There are various ways to assign how the variance increases as one moves away from a known point as in Fig. 7.3.

The above is a brief explanation of Bayesian search, but what the beginners may be wondering is why this is called 'Bayesian'. To clear up the haze on this point, we formulate it in a little more detail below.

In Fig. 7.3a, there are two known points at k-step, between which an interpolated estimation $\mu_k(x)$ (solid line) is given as the value of evaluation function accompanied with the uncertainty $\sigma_k(x)$ (dotted lines). The parameters to dominate the situation at k-step are then denoted as

$$\boldsymbol{\theta}_k \sim \begin{pmatrix} \mu_k(x) \\ \sigma_k(x) \end{pmatrix},\tag{7.5}$$

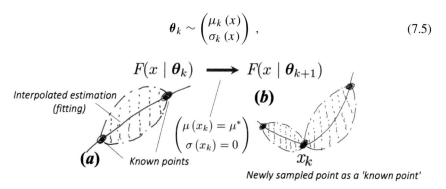

Fig. 7.3 Schematic explanation of Bayesian search. The evaluation function value $\mu(x)$ is estimated by interpolating the known points (solid line). The interpolation is accompanied with the errorbar σ (dotted line) so that it may increase as one moves away from a known point

and the $A(x)$ appeared above determines the next sampling point selected at k-step, shown as x_k in Fig. 7.3b. Let us conceptually denote the situation of the search space at k-step as $F(x \mid \boldsymbol{\theta}_k)$.

Getting x_k, the parameters μ and σ should be updated so that

$$\mu_{k+1}(x_k) = \mu^* \quad , \quad \sigma_{k+1}(x_k) = 0 , \tag{7.6}$$

can be satisfied [3]   [Fig. 7.3b]. By this update,

$$\boldsymbol{\theta}_k \sim \begin{pmatrix} \mu_k(x) \\ \sigma_k(x) \end{pmatrix} \rightarrow \begin{pmatrix} \mu_{k+1}(x) \\ \sigma_{k+1}(x) \end{pmatrix} \sim \boldsymbol{\theta}_{k+1} , \tag{7.7}$$

the next sampling point x_{k+1} is determined by the updated situation $F(x \mid \boldsymbol{\theta}_{k+1})$.

As a general situation, let us suppose an event x which occurs under some stochastic rule governed by the parameter $\boldsymbol{\theta}$ (e.g., the mean μ, the variance σ etc.) Bayesian estimation is used to estimate unknown $\boldsymbol{\theta}$ based on the actual x observed [9]. In our problem, the situation is indeed 'how to estimate $\boldsymbol{\theta}_{k+1}$' updated from $\boldsymbol{\theta}_k$) when x_k is actually obtained. The name 'Bayesian' in this subsection comes from the fact that the update $\boldsymbol{\theta}_k \rightarrow \boldsymbol{\theta}_{k+1}$ is implemented by applying Bayes' estimation rule (Sect. 13.1).

What confuses beginners is 'which distribution is updated by the Bayesian framework?'

Since the mission of the search is to determine the next sampling point x_{k+1}, one would tend to misunderstand that the x-updating is subject to the Bayesian updating. But it is actually updated as

$$x_{k+1} = \arg \max_{x} A_{k+1}(x) , \quad A_{k+1}(x) = \mu_{k+1}(x) + \lambda \cdot \sigma_{k+1}(x) ,$$

without any taste of Bayesian. What is updated by the Bayesian framework is actually the parameter $\boldsymbol{\theta}_k$ rather than x_{k+1}. Using the term 'distribution', the target of Bayesian updating is not 'the distribution in the search space for x' but 'the distribution in the parameter space for $\boldsymbol{\theta}$'. The terminology of "Bayesian 'search'" might be improper because it may cause the misunderstanding that the search space for x is handled by the Bayesian framework.

We note that for $A_{k+1}(x)$, the above expression is just an example, but there are several different schemes how to compose it not only by the linear combination but also by more sophisticated functional forms [9].

---

[3] At the new sampling point x_k, the value of the evaluation function has actually determined as μ^* without any uncertainty, as such $\sigma = 0$.

7.3 Regression Using Descriptors

In Sect. 7.1.3, we introduced the concept of a Y ($\mathbf{X}$) regression relation, in which Y denotes materials properties being associated with a multi-component descriptor $\mathbf{X}$. In this section, we explain how to construct such a regression relation.

7.3.1 Binary Tree Regression

When one gets a set of given data as described in Fig. 7.4, one may draw a straight line of linear regression as in (a), based on the knowledge learnt in undergraduate. On the other hand, it is also possible to view the data as in (b). In this case, the data is described as

$$y = \begin{cases} y_1 & (x < x_1) \\ y_2 & (x_1 < x < x_2) \\ y_3 & (x_2 < x < x_3) \\ \vdots & \end{cases} \tag{7.8}$$

For a given new data x, the corresponding y is estimated based on this case separation, being another possible form of regression. In this case, however, the number of parameters required to realize the regression gets large compared to the parametrization in linear regression (two parameters, intercept and slope), seeming quite inefficient. Furthermore, how to define the interval $\{x_j\}$ has too much room for arbitrariness, seeming to be troublesome. However, with modern data processing capabilities, it is possible to comprehend data like (b) at high speed, called **binary tree regression**[4] [10]. One may understand that, in this regression method, the regression relationship is not restricted to linear and can be described flexibly.

Figure 7.4b was the case of a 1-variable descriptor, $y = y(x)$. In the case of a y-value corresponding to a 2-variable descriptor, $y = y(x_1, x_2)$, we can partition the (x_1, x_2)-plane as in Fig. 7.5a and the y-values, $y_1, y_2, y_3, y_4, \cdots$, are assigned to each parcel. This assignment situation can be represented by a two-layered binary-tree as shown in Fig. 7.5b.

Even if the number of descriptor components increases to 3-variable or 4-variable $\cdots$, the same idea can be applied, in principle, to regression representation of data by dividing the multidimensional space into diced parcels (as hyper cubic). In this case, the tree branching becomes more multi-layered and looks like a tree, which is why it is called a binary-tree. In the example in Fig. 7.5a, $\mathbf{x} = (x_1, x_2)$ is divided into four regions, but if we make the region division more detailed, we can capture more

---

[4] Given a new data point x, the regression determines which interval it belongs to such as "'$x > x_5$' or '$x < x_5$'?" and "Then, '$x > x_3$' or '$x < x_3$'?", by asking a series of questions in a tree-like fashion. The method is then named as 'decision tree' or 'binary tree'.

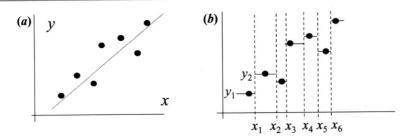

Fig. 7.4 How to view the given data sequence. Classically, the data is comprehended by a linear regression as in (**a**), but in modern data science, it is comprehended as in (**b**) with the help of processing power

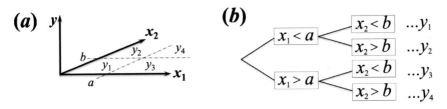

Fig. 7.5 The dependence on a 2-variable descriptor, $y = y(x_1, x_2)$, is represented by a two-layered binary-tree

diverse and detailed dependencies as a function. However, the more grid is taken too finely, it would deviate from the regression philosophy to describe the dependence with a small number of parameters.

Let us consider, for example, 'x_1 = occupation, x_2 = gender, x_3 = country of residence, ...' to estimate the annual income y by repeating the conditional branching as shown in Fig. 7.5b. You can easily imagine that rather than sorting x_1, x_2, x_3 in that order without any consideration, it would be better to sort the whole thing first with x_3, then classify it with x_1 followed by other descriptors. By doing this, we can omit irrelevant x_j to describe y to achieve the reduction of parameters in the regression. As seen in this example, there is not only one way to take a binary tree (what is the first branch and what is the second branch), but also there are many possible ways. Also, depending on what you choose as the first branch, you can achieve '(better expression with fewer branches) = (regression with fewer parameters)'.

The efficient choice of the first branch depends on what x_j has the most influence on the y values. In the entire descriptor space, there is a situation where y values are most sensitive to x_3 in some region, but in another region, the dependence on x_2 is rather strong, for instance. As such, the proper choice for the first branch depends on the region of interest. Based on this observation, we can take an idea to adopt each regression in parallel using several different branching orders, and to complement descriptiveness by using those regression values in a complementary manner (**ensemble learning**) [10]. The multiple possibilities of how to take the branching order, however, gets enormous easily as the number of descriptors increases. So, the easiest strategy on which branching order to be taken would be the random choice, called **random forest** [10].[5] In Sect. 7.1.2, we saw that random sampling is a starting point toward more efficient sampling strategies (particle swarm, Bayesian, etc.). Similarly, the random forest strategy has been improved toward more efficient way to pick up relevant branching orders (called boosted trees for random forests) [10]. Such strategies include 'gradient boosting' as a representative improvement.

7.3.2 Regression Using Neuralnetwork

Consider a '2-input, 1-output device' as depicted in Fig. 7.6b that mimics the function of a neuron. The output value takes $y = 1$ (on) or $y = 0$ (off), as determined by the input values as

$$y(x_1, x_2) = \begin{cases} 0 & (w_1 \cdot x_1 + w_2 \cdot x_2 < b) \\ 1 & (w_1 \cdot x_1 + w_2 \cdot x_2 > b) \end{cases} . \tag{7.9}$$

The condition to split y-values into 0 or 1 can be expressed by a straight line in the (x_1, x_2)-plane, as depicted in Fig. 7.6a.

Next, consider a 2-fold construction as shown in the dotted box in Fig. 7.7b, formed by two basic elements (black circles). Then, the values taken by (y_1, y_2) are determined by (x_1, x_2) as shown in Fig. 7.7a as represented by the domains separated by straight lines in the (x_1, x_2)-plane.

Using the (y_1, y_2) to further construct z as shown outside the dotted box in Fig. 7.7b, the value of $z = 0$ or 1 is assigned on the domains of the (x_1, x_2)-plane in Fig. 7.7a. By increasing the number of x_j, the function gets to be extended from a 2-dimensional to a multidimensional function. By increasing the number of y_j, the domains in $(\{x_j\})$-plane get to a finer division. As such, the complex neural network as shown in Fig. 7.8a can be understood to be such a more detailed expression of a multidimensional with finer resolution.

As long as we use elements with Eq. (7.9), the domain segmentation can only be represented by straight lines.[6] The Eq. (7.9) can be written using the step function $u(x)$ as

$$y(x_1, x_2) = u(w_1 \cdot x_1 + w_2 \cdot x_2 - b) . \tag{7.10}$$

Its discontinuous jump between 0 and 1 corresponds to the fact that (x_1, x_2)-plane delimited by a straight line.

By replacing the step function $u(x)$ with $h(x)$ whose value varies continuously as shown in Fig. 7.8b, we can make it into $y(x_1, x_2) = h(w_1 \cdot x_1 + w_2 \cdot x_2 - b)$, which gives the flexibility of the region segmentation to bend in response to the edge collapse of $u(x)$. The output value y can also have continuous values between 0 and 1 to represent continuous changes across regions.

---

[5] The word 'forest is used because multiple decision 'trees' are taken.
[6] Such an element is called the **perceptron** [10].

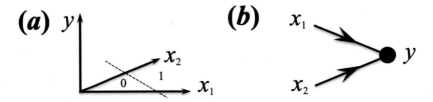

Fig. 7.6 The basic element to form a neural network [panel (**b**)]. The 2-inputs and 1-output are comprehended as the value y (x_1, x_2) given on the (x_1, x_2)-plane [panel (**a**)]

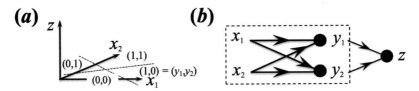

Fig. 7.7 Two basic elements of the neural network are superimposed [inside the dotted box in panel (**b**)], the 2-inputs and 2-outputs are comprehended as the values (y_1, y_2) on the (x_1, x_2)-plane [panel (**a**)]

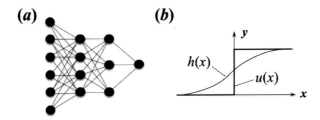

Fig. 7.8 An example of neural network configuration [panel (**a**)] and activation function [panel (**b**)]. u (x) stands for the step function and h (x) is the activation function

By superimposing a number of elements with h (x) instead of u (x) [as shown in Fig. 7.8a] to form an N-input 1-output network, we can thus construct a flexible regression of N-variable functions. Such a network includes many adjustmentable parameters, $\left(w_1^{(\alpha=1,\cdots,N)}, w_2^{(\alpha)}, b^{(\alpha)} \right)$. By adjusting these parameters, we can realize flexible regressions. It is actually known that such neural networks can represent arbitrary functions, called **universality theorem** [15].

The h (x) is called **activation function** [10,17]. There are many ways how to construct a shape of network and how to choose the functional form for the activation function. It is one of the missions for methodological experts of the neural network to consider how to efficiently optimize a huge number of adjustable parameters, how to choose the network configuration to achieve the best efficiency under the feasibility of the optimization.

7.3.3 Evaluating and Improving Regressions

Whether it is binary tree regression or neural network regression, one can conveniently utilize well-developed tools implemented on Python to perform data regression nowadays [16].

As the indicators to evaluate regression performance, RMSE (Root Mean Squared Error), MAE (Mean Absolute Error), and the coefficient of determination R^2 are commonly used [17]. While RMSE measures the deviation from the mean by its squared quantity, MAE measures it by its absolute value. The name **Robust Estimation** [18] is sometimes used for strategies adopting the absolute values instead of squared values, because if large outliers with only a few points are measured by the squared weights, the forecast itself would be greatly affected by the huge squared quantities and become blurred.

For both RMSE and MAE, their magnitude scales with the range of the target y value. Consider a case, for example, that the error in measuring the melting point of Tungsten ($\sim$3,300 °C) marks $\pm$ 5 °C, while the error for the boiling point of water ($\sim$100 °C) gets to $\pm$ 3 °C. For this case, we cannot say that the measuring error is smaller for water. The former has a ratio of $5/3300 = 0.002$, while the latter has a ratio of $3/100 = 0.03$, which means that the former (Tungsten) has measured with far less error. As such, it is important to compare the qualities under the **normalization** by each size of the quantity. The R^2 is the normalized measure of the square error (it originally scales to the range taken by the y value) so that it may take values in the [0, 1] range, allowing cross-comparison over different statistics. R^2 is normalized so that $R^2 = 1$ when all data are on the regression curve and agree.

In recent practice, one can prepare descriptors by using descriptor generation tools as described in Sect. 7.1.1 [4]. Leaving aside the question of whether the descriptors make sense as explanatory variables, the tool generates several hundred descriptors without any consideration for now. The first step is to run a regression with all of them as descriptors, and then reduce the number of descriptors by referring to the results of the regression.[7] Section 13.2, we have given a brief explanation on **norm regularization** as one of the methods to reduce the number of descriptors. The norm regularization includes **LASSO**,[8] which becomes popular in Materials Informatics.

In connection with this technique, semantically curious phrases such as 'regularization', 'trimming of descriptors', and 'sparse modeling' [19] frequently appear and may confuse beginners. The 'trimming' is not so strange word as it has the sense of reducing the number of descriptors. Why this is called **regularization** needs a little explanation for beginners. When N data are explained by N descriptors, the excess or deficiency of information is balanced (regular condition), but when the number of descriptors is small or large, the excess or deficiency of information arises and becomes an ill-posed problem. In the current context, the wording is to cut down on

---

[7] Reducing the number of descriptors to narrow down the relevant ones is a common practice known as the factor analysis in statistical survey for social sciences, psychology, etc., where a classic method based on the covariance and correlation coefficients are used.

[8] LASSO (Least Absolute Shrinkage and Selection Operator).

the number of descriptors, which are too many, and to bring in a regular situation, i.e., regularization. For the term 'sparsity', it appears in the context that the phenomenon should be described by a very small number of descriptors (**sparsity**). Considering an overfitted regression with too many descriptors, essentially relevant descriptors are 'sparsely' found among unnecessary ones, that is what the wording means.

In general, when one tries to explain data using too many parameters (like descriptors), one would end up with a regression curve that unnaturally traces all the given data points (**over-training**) [10]. Such a *special* regression is never capable to capture the general trend which is also valid for another sampling of data. In over-trained data fitting, the regression curve is forced to follow all the data points, and some parameters take very large values, such as plus or minus tens or minus hundreds of thousands, to produce an unnaturally winding curve.[9] It is not possible to make sense of descriptors with too unrealistically large values. The idea of norm regularization (Sect. 13.2) is to monitor the absolute value (norm) of each descriptor and trim off the ones that are too large in its magnitude.

In the case of the binary tree regression [Sect. 7.3.1, Fig. 7.4b], the regression is not like that accompanied with expanding coefficients at all. Therefore, the norm regularization strategy, i.e., eliminating those with large norms, is not applicable. Binary tree regression is a regression implementation based on the concept of data classification by conditional branching. If the branching is made too complicated with a large number of branching conditions, the given data can be classified perfectly, but the performance of interpolation for unknown data will be inferior due to overlearning. If more efficient classification can be achieved with as few branches as possible, the number of descriptors for conditional branching can be reduced. This corresponds to regularization for binary tree regression.

As mentioned at the end of Sect. 7.3.1, the efficiency of the binary tree regression depends on the choice of which descriptor to be employed as the first branch and then which for the second branch. Such implementations are provided that allows these choices to be made automatically while utilizing a quantitative measure of classification efficiency (explained below). In binary tree regression, all given data points are classified into various segment labeled by y_j in Fig. 7.5a. During the branching until the final classification is completed, data to be finally sorted into different y_j would coexist for the time being in the same group. This is a situation where finite *impurity* occurs within the group. As the branching proceeds, the impurity decreases as the classification proceeds. There is a decrease in impurity (called information gain) when moving from the k-order branching to the $(k + 1)$-order branching, and the descriptors to be adopted for the next branching condition are selected so that this decrease is as large as possible. For the measure how much the impurity gets decreased, Gini-impurity (that has been used in Economics), and entropy are known as representative indices [10].

---

[9] When trying to express a largely winding curve with a power-expanded function, the expansion coefficients of the higher order terms take very large values with alternating sign.

7.4 Structural Search

7.4.1 Structural Search Using Regression

By using the framework described in Sect. 7.3, we can construct an evaluation function obtained by regression. As explained in Sect. 7.1.3, we do not perform 'inverse estimation', but the forward estimations repeatedly so that the next sampling points in the descriptor space can satisfy as possible

$$X^* = \arg \max_X Y(X) , \qquad (7.11)$$

by using the search methods described in Sect. 7.2 [Bayesian/particle swarm/genetic algorithm].

As an example, when the authors performed the search over BCS superconductors of hydrides for higher transition temperature (Y) [20], we started with picking up data for regression from public databases and original papers. Next, we prepared about 300 descriptors $(X_j \sim \mathbf{X})$ by using tools openly available in Chemoinformatics (e.g. XenonPy [1]) to characterize chemical composition, applied pressure, crystal structure etc. Then, we constructed a binary tree regression to describe $(\mathbf{X} \to Y)$ using the open-source machine learning library 'Scikit-learn' [16]. We further trimmed the number of descriptors as explained in Sect. 7.3.3 to achieve a high R^2 just with 84 descriptors [4].

In the above example [4], a genetic algorithm is used for structure exploration, but we note that the algorithm is not used to perform Eq. (7.11). The optimization of Eq. (7.11) is performed in terms of the chemical composition, not of the crystal structure. Once the optimized composition X^* has been determined, we used a genetic algorithm to get possible structures for an optimized composition.

In that algorithm, The initial structures for X^* are created in hundreds of random atomic configurations and each energy is evaluated. Among all structures, only those achieving lower energy are selected in the parent generation. The selected parents are used to create the structures in the child generation by genetic operations as described in Fig. 7.9. This is repeated several times to achieve a convergence in the sense that the members of structure achieving lower energy get to be almost the same in the reputation. Then the final members are employed as the optimized structures. In Sect. 7.2.2, we introduced genetic algorithms in a general context to update descriptors by the genetic operation. In the present context, on the other hand, the crystal structures are directly handled by the genetic operation to generate new structures, as in Fig. 7.9 [13].

7.4.2 How ab Initio Calculation Used

As mentioned at the end of Sect. 7.1.3, ab initio calculations are not necessary an essential piece in either regression construction or data exploration. In fact, in the example in the previous subsection, ab initio calculations are not used in the regression construction and data search. Ab initio calculation appears *after* the composition

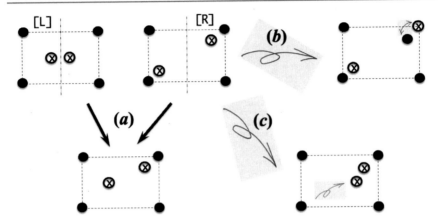

Fig. 7.9 Genetic manipulation of structure. The structure (**a**) is generated by crossover of the structures of [L] and [R]. The structure (**b**) is generated by permutation of the lower rightmost • and x in [R]. The structure (c) is generated by performing a mutation on the position of 'x' in the cell of [R]

X^* has been determined in the data search, where candidate structures are generated by a genetic algorithm.[10] Ab initio calculations are also used to check whether the finally predicted structure is really energetically stable. Without this, the prediction would be pie in the sky. In this section, we explain how ab initio calculations are used in the context of the structural search in practice.

For a predicted structure, we first inspect its thermodynamic stability, i.e., whether the composition in question can exist without decomposing into smaller components. For example, whether $LaYH_{12}$ can exist at least at zero temperature without decomposing into $LaYH_{12} \rightarrow LaH_6 + YH_6$. This inspection is performed by an ab initio calculation including structure optimization (Sect. 5.1.2) for each compound, evaluating the enthalpy $H(P)$ from the ground state energy and the optimized volume.[11] As shown in Fig. 7.10, we can predict which structure is thermodynamically most stable in which pressure range, by plotting the enthalpy $H(p)$ against several candidate structures predicted by the genetic algorithm. By comparing with the sum of enthalpies of decomposition products, we can also predict the stability against decomposition. In the example in Fig. 7.10, the $R\bar{3}c$ ($Cmmm$) structure is stable at lower (higher) pressure, and both are predicted to undergo a structural phase transition upon pressure application without decomposition [22].

Then we check the **dynamical stability** of the candidate structure using ab initio phonon calculations [23]. The enthalpy comparison above compares the energy

---

[10] Although this in itself is a structural 'search' for a given composition, it is not a search to get a solution X^* in descriptor space. Do not be confused by the term 'search' in this aspect.

[11] The enthalpy $H(P) = E(V) + p \cdot V$ can be calculated given the volume dependence of energy $E(V)$. Recent ab initio packages have a function to calculate it by using the numerical gradient of $p \sim -\partial E / \partial V$ or the equation of state (e.g. Birch–Murnaghan equation of state etc. [24]) which gives $p = p(V)$.

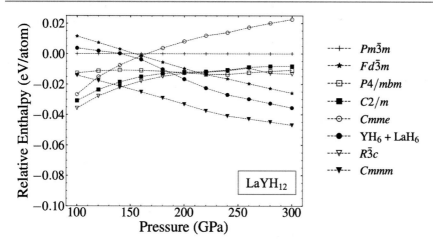

Fig. 7.10 Enthalpy comparison over crystal structures predicted by a genetic algorithm for a superconductor LaYH$_{12}$ [22]. The $R\bar{3}c$ ($Cmmm$) structure is stable at lower (higher) pressure, and both are predicted to undergo a structural phase transition upon pressure application without decomposition

values evaluated for a fixed geometry of atomic positions of an assumed crystal structure. If the energy gets stabilized more when the atomic position deviates even slightly from the fixed position, it means that the assumed structure itself cannot exist stably and collapses into another structure. If the assumed geometry is stable, then the energy change upon adding a small displacement u to the atomic position should rise to a leading term with a quadratic term,

$$\Delta E \sim A \cdot u^2 + \cdots , \tag{7.12}$$

where A is positive.[12] If we replace The energy uplift due to A is equivalently described by the restoring force of the spring constant $F \sim k \cdot u$ with $k = M\omega^2$ as

$$\Delta E \sim \frac{1}{2}k \cdot u^2 = \frac{1}{2}M\omega^2 \cdot u^2 , \tag{7.13}$$

leading to

$$\omega \sim \sqrt{\frac{2A}{M}} . \tag{7.14}$$

This vibration frequency ω is called the phonon frequency.[13]

---

[12] It follows from the logic that ΔE must be positive regardless u takes positive or negative, so the leading term should be even power.

[13] u is the amplitude of a normal mode of a crystal lattice and M is the converted mass of the vibration mode.

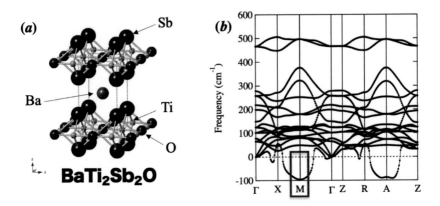

Fig. 7.11 A specific example of crystal structure stability evaluated by phonon calculations [25]. Panel **(b)** shows the phonon dispersion diagram for the crystal structure [panel **(a)**]. Looking at M point surrounded by the frame, we can see that imaginary phonon frequencies appears

When A is negative, it means that ΔE gets to be more stabilized with increasing the distortion u, corresponding to the assumed structure being unstable. For the negative A, the corresponding phonon frequency ω becomes imaginary from Eq. (7.14). As such, when a phonon calculation is performed to get imaginary phonon frequency (it is plotted as negative values in the phonon dispersion diagram), it implies that the assumed structure will spontaneously be deformed. In other words, the crystal structure cannot exist dynamical stably.

A specific example is shown in Fig. 7.11 [25]. The phonon dispersion diagram for the crystal structure [panel (a)] is shown in panel (b). The horizontal axis of the dispersion diagram corresponds to the wavenumber vector (3-dimensional), which is made into 1-dimensional axis by taking the 1-dimensional path connecting between the characteristic points in the 3-dimensional reciprocal lattice of the crystal to be considered (Sect. 3.6). The wavenumber specifies the way how the mode changes when it advances by one lattice period, as explained in Fig. 11.3. For the atomic displacement described by the wavenumber, $k = \pi$ means such displacement taking the opposite direction for every 1 site shift (the displacement getting back to the same direction with the original amplitude by every 2 sites shift), while $k = 2\pi/3$ means the displacement to take the same direction with the original amplitude by every 3 sites shift. In the 3-dimensional case, the displacement patterns represented by each **k** are more complex, but in any case, each point on the horizontal axis of the dispersion diagram corresponds to a different deformation pattern of the crystal lattice. For each deformation pattern (each point in the horizontal axis), we can plot the phonon frequency of Eq. (7.14), which corresponds to the energy uplift as the vertical axis quantity. Connecting the vertical axis quantities with continuous curves provides the phonon dispersion diagram.

In Fig. 7.11b, the M and A points are highlighted by a frame showing negative phonon frequencies. It is the convention to display the imaginary modes as negative values. This result therefore means that a vibrational displacement corresponding to the M point leads to further stabilization in energy, namely that the originally

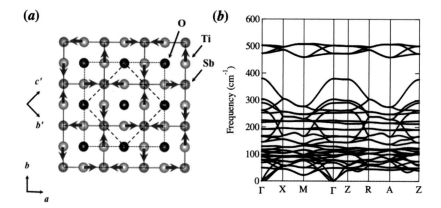

Fig. 7.12 The panel (**a**) shows the spatial pattern of the deformation corresponding to $\mathbf{k}_M$ at the M point in the panel (**b**) of Fig. 7.11. Along with this displacement, the structure is further relaxed until the energy gets to be the most stable, leading to the dispersion shown as panel (**b**), where the imaginary modes no longer occur

assumed structure is unstable toward spontaneous deformation. The spatial pattern of the deformation corresponding to $\mathbf{k}_M$ at the M point can be illustrated from the output file of an ab initio package, as shown in Fig. 7.12a. The emergence of an imaginary mode at the M point then means that there emerges the spontaneous deformation as the displacement in Fig. 7.12a. The structure should then be relaxed along this displacement until the energy gets to be the most stable. The dispersion diagram for the relaxed structure is shown in Fig. 7.12b. Since there are no more imaginary modes here, we can conclude that the structure is now dynamically stable.

Returning to the context of structural exploration, it is then necessary to check ab initio phonon calculations to get the phonon dispersion, seeing whether no imaginary modes appear to ensure dynamic stability, namely the predicted structure for the desired properties obtained by data scientific schemes can really exist stably. The dynamic stability check by phonon calculation can be used not only for structural exploration, but also for the possibility of experimental synthesis for the question, 'Is such a synthesis possible? (Is the structure of the product dynamically stable?)', 'Why doesn't such synthesis occur as expected? (Is it due to the instability of the product?)'.

References

1. https://xenonpy.readthedocs.io/en/latest/# (URL confirmed on 2022.11)
2. https://daylight.com/dayhtml/doc/theory/theory.smiles.html (URL confirmed on 2022.11)
3. "Bayesian Molecular Design with A Chemical Language Model", H. Ikebata, K. Hongo, T. Isomura, R. Maezono, and R. Yoshida, J. Comput. Aided Mol. Des. 31, 379 (2017). https://doi. org/10.1007/s10822-016-0008-z
4. "High-Tc ternary metal hydride, YKH$_{12}$ and LaKH$_{12}$, discovered by machine learning", P. Song, H. Zhufeng, P.B. de Castro, K. Nakano, K. Hongo, Y. Takano, R. Maezono, http://arxiv. org/abs/2103.00193

5. "*Ab initio* random structure searching", Chris J. Pickard and R.J. Needs, J. Phys.: Condens. Matter 23, 053201 (2011). https://doi.org/10.1088/0953-8984/23/5/053201
6. "An Introduction to Metaheuristics for Optimization", Bastien Chopard, Marco Tomassini, Springer (2018/11/13) ISBN-13:978-3319930725
7. "Particle swarm optimization for single objective continuous space problems: a review", M. R. Bonyadi, Z. Michalewicz, Evolutionary Computation. 25, 1 (2017). https://doi.org/10.1162/EVCO_r_00180
8. "Introduction To Genetic Algorithms For Scientists And Engineers", David A Coley, World Scientific (1999/1/29) ISBN-13:978-9810236021
9. "Bayesian Optimization and Data Science", Francesco Archetti, Antonio Candelieri, Springer (2019/10/7) ISBN-13:978-3030244934
10. "The Elements of Statistical Learning: Data Mining, Inference, and Prediction", Trevor Hastie, Robert Tibshirani, Jerome Friedman, Springer (2009/3/1) ISBN-13:978-0387848570
11. "Mathematical Methods for Physics and Engineering", Ken F. Riley, Mike P. Hobson, Stephen J. Bence, (Third Edition), Cambridge University Press (2006/6/8) ISBN-13:978-0521683395
12. http://www.calypso.cn (URL confirmed on 2022.11)
13. https://uspex-team.org/en (URL confirmed on 2022.11)
14. http://xtalopt.github.io (URL confirmed on 2022.11)
15. "Approximation by superpositions of a sigmoidal function", G. Cybenko, Mathematics of control, signals, and systems 2, 303 (1989). https://doi.org/10.1007/BF02551274
16. https://scikit-learn.org/stable/ (URL confirmed on 2022.11)
17. "Hands-on Machine Learning With Scikit-learn, Keras, and Tensorflow: Concepts, Tools, and Techniques to Build Intelligent Systems", Aurelien Geron, Oreilly & Associates Inc (2022/11/15) ISBN-13:978-1098125974
18. "A Student's Guide to Data and Error Analysis", Herman J.C. Berendsen, Cambridge University Press (2011/4/7) ISBN-13:978-0521119405
19. "Sparse Modeling: Theory, Algorithms, and Applications", Irina Rish, Genady Grabarnik, CRC Press (2014/12/1) ISBN-13:978-1439828694
20. "Potential high-Tc superconductivity in YCeH20 and LaCeH20 under pressure", P. Song, H. Zhufeng, K. Nakano, K. Hongo, R. Maezono, Mater. Today Phys. 28, 100873 (2022). https://doi.org/10.1016/j.mtphys.2022.100873
21. https://phonopy.github.io/phonopy (URL confirmed on 2022.11)
22. "High T_c superconducting hydrides formed by the cage structures LaH24 and YH24 as basic blocks", P. Song, H. Zhufeng, P.B. de Castro, K. Nakano, K. Hongo, Y. Takano, R. Maezono, Chem. Mater. 33, 9501 (2021). https://doi.org/10.1021/acs.chemmater.1c02371
23. "Density Functional Theory: A Practical Introduction", David S. Sholl, Janice A. Steckel, Wiley-Interscience (2009/4/13), ISBN-13:978-0470373170.
24. https://www.quantum-espresso.org (URL confirmed on 2022.11)
25. "Phonon dispersions and Fermi surfaces nesting explaining the variety of charge ordering in titanium-oxypnictides superconductors", K. Nakano, K. Hongo, and R. Maezono, Sci. Rep. 6, 29661 (2016) https://doi.org/10.1038/srep29661

Tips in Project Management

8

Abstract

This chapter will discuss some of the tips in practical situations in simulation studies, especially for situations where simulation is incorporated by groups based on experimental research. Based on the authors' experience, I will also discuss a few somewhat crude issues I learned. Analysis on what kind of factors hinder the motivation for the collaborations is provided, those are found in unexpected places.

8.1 How to Drive the Collaboration Effectively

This book basically assumes such target audience whose main task is to conduct experiments mainly without simulation research infrastructure initially. For these audience, we would like to provide tips on how to build the a scheme effectively enabling to conduct ab initio simulations **on their own** in the near future. Our experience shows that the successful cases always assumed such collaboration where the experimental side aimed to get to be able to run the simulations on their own, which is found to be the key to success. We have actually been developing a teaching program to support this point, which is the background for writing this book.

8.1.1 The Age of Experimental Practitioners Running Their Own Calculations

Collaborations between simulation and experimental groups are never the novel concept, that has been around for a long time. However, for the most part, it has taken the form where the simulations are handled by the members in simulation group side. Such collaborations are actually found not to proceed smoothly. Main reason for this is the lack of manpower. The members in simulation group side do not have such strong **the passion as the primary player**, especially in the academic

© The Author(s), under exclusive license to Springer Nature Singapore Pte Ltd. 2023   195
R. Maezono, *Ab initio Calculation Tutorial*,
https://doi.org/10.1007/978-981-99-0919-3_8

situation, where the members are usually engaged in research on methodologies primary. For such a work given in the collaborations, it is natural for them to treat it as 'secondary priority' because they cannot be the first author. Regarding this situation as the market, there are commercial entities that specialize in contracted simulation analysis, but the price for the analysis is generally very expensive.

Since ab initio methods have become sufficiently mature, experimentalists themselves are now able to perform calculations by themselves with a little learning. The number of such capable experimentalists with simulation skills is actually increasing internationally in recent year, which gets to a sobering threat to experimentalist in the world-wide competition. As such, there increases such cases recently that experimentalists (mainly graduate students belonging to experimental groups) take short visit to my group to learn computational methods. Once the students in the experimental group start learning ab initio analysis, the project is accelerated significantly. They have the **passion as the primary player** for publicize their own samples or measurements that they synthesized or performed by their own with their pride. They are naturally urged to construct their scenario for their achievements assisted by ab initio analysis. This clear objective to conduct theoretical analysis to prove it can be **the strong driving force** to go into the modeling process to find out what is the essence of the problem the experimentalists is working.

The collaboration that we consider in this chapter is not a 'leave it to you'-type (outsourced form), but rather a education-collaboration type where a member from experimental group is aiming at acquiring the simulation but for the time being assisted by the collaborator (calculation experts). The general flow of the process would be 'literature review $\rightarrow$ consultation with experts $\rightarrow$ tracing of preceding calculations $\rightarrow$ gradual transition to practical work', as describe in the following sections in order. In addition, we will also consider the situations where a organizer of an experimental group (supervisor, group leader etc.) supervises his or her team members to gradually set up a scheme for actual simulation work.

In the above, the process of 'consultation with experts' requires another tip in selecting the right person to inquire. In some countries, there are quite a few 'basic-oriented researchers' who are not so interested in practical applications. In some community, they have a sort of pride that they have contributed to DFT or molecular orbital (MO) method providing key fundamental methodologies since the time when DFT or MO was first established, so the more prestigious a laboratory becomes, the more specialists are interested only in pioneering sharp methodologies and show little interest in practical applications. I've heard from several industrial practitioners that they once tried to talk to such prestigious theoretical groups ending up with the comments like 'your problem is too difficult, because of this factor, this mechanism, ... ', just explaining the long story why it is difficult, and they leave.

Since the author myself is another theoretical researcher in the methodology of stochastic many-body electronic theory, I am not unaware of this feeling. However, we note that coping with solving complex phenomena that are 'known to be complex' and how to model them to break down with feasible tools[1] has been originally an important mission of scientific research (not only natural science, but also science in general including social science etc.). Today, methodologies have developed to such a high level that only the aspect of pursuing novel methodology itself would be too emphasized as a mission of Science (especially in such a country where a history of proud contributions to the establishment of methodologies, coupled with traditionally sophisticated craftsmanship, this tendency may be strengthen). Depending on the country or time period, however, the ability of calculus or linear algebra would be regarded just as a skill to use tools like a screwdriver or wrench, being a mission of technical education. There would be another academic culture that emphasizes the importance to develop insights into complex problems rather than tools themselves as the mission of higher education. It is actually surprisingly difficult to find a theoretical expert who is happy to work with you on modeling with such feeling above.

8.1.2 What Makes it Different from a Simple Analytical Collaboration

In the usual case of collaboration in experimental fields, the standard picture would be 'a sample synthesized by Prof. A's group, X-ray structure analysis by Prof. B's group, and mass spectrometry by Prof. C's group', so we often receive inquiries such as 'How long will it take to make simulation analysis on this sample?'. However, the time required for the 'analysis' is usually difficult to be estimated. As explained in Sect. 1.3.2, ab initio analysis is not a 'whole simulation as it is', so it is quite different from the picture as 'applying a received sample to a measurement'. The ab initio analysis requires **modeling**, to reduce the issues to a computable model that can capture the essense of the interested behaviour of your samples. It is impossible to simulate the tremendously complex problems of mixing ratios, amorphous properties, defects, and interfaces as they are.

The modeling itself is the main part of the collaboration, as such it is impossible to estimate how long it will take before it gets to the stage for the calculations on computers. Once we get into the computation, we can surely estimate how long it will take in terms of CPU time or memory capacities. Once the calculations are completed, however, the results (just energies of charge distributions) do not immediately tell us what we want to know, as mentioned in Sect. 1.2.1. Rather time is required for insight into how to interpret the calculation results to relate with the phenomenon or behavior we want to understand. Often, the result of the calculation ends up with a negative result, such as 'required resolution to distinguish between hypothesis is not possible with the current computational power'.

This is the reason why the simulation side are sometimes puzzled when asked how long the analysis will take. If the required modeling has already been done

---

[1] In our case, that is the tool that can only calculate ground state energies at zero-temperature.

by the experimental side asking to run the simulation with this geometry on your supercomputer, it would be the similar situation with that 'I will provide the sample and ask you to perform X-ray structural analysis. Could I ask how long it will take?'. However, it would be quite unlikely that 'an experimental practitioner will bring in a collaboration after he/she has completed the psi-zero modeling'.

8.1.3   How to do a Literature Search

'How to model the practical problems by ab initio methods' as explained in Sect. 1.3.2 is a challenging research problem in itself, and is not something that a non-major practitioner can easily come up with on his/her own from scratch. One may think that 'theoretical researchers should do it, since it is their duty as scientists to think about such modeling', but they do not have enough hands or brains to cover all modeling for the many phenomena in materials science (if one establishes even one such modeling strategy, it will be a milestone-achievement for him/her as theoretician). But don't be so pessimistic. This book assumes the situation where a practitioner just starts to consider learning ab initio technique stimulated by knowing that the analysis is frequently used in his/her research topic. In such situations, there is usually some prior researches.

For your interesting topics of properties, just try Google search 'e.g., photocatalyst' + 'DFT/ab initio/first principles', leading to find most of the previous studies by which you can get the idea how to model your problem captured by ab initio technique. In this case, it is important to know where to focus your attention in the literature you find.

A reader who has read this book up to this point should be able to understand the wording of the following,

- (a) After all, 'which quantity' is evaluated by ab initio calculation to argue the problem of interest,
- (b) How the discussion is constructed based on the calculated results.

It is important to 'steal the strategy' from the previous studies. In addition to the strategic aspect, one can also get technical aspects, such as details of computational specifications from Supporting Information.[2]

If you are a complete beginner, it may be difficult to fully understand (a) and (b) above. However, since you are a researcher, you may be able to identify where (a) and (b) is written in a paper. Once you identify the points, then bring it to the simulation specialist to make the project go faster. As mentioned above, the modeling of the problems in materials science is diverse and constantly evolving, as such even experts in simulation methods cannot fully comprehend it. However, if there is a

---

[2] In recent academic papers, details of calculations or experiments are now provided as Supporting Information.

prior paper to be handed to the experts, they should be able to understand how about (a) and (b) for your problem treated in the preceding paper. The experts will be able to read about the difficulty/feasibility, software requirements, and so on, and this information will make the research plan much more realistic to be realized.

! Attention

It may be obvious to practitioners, but it is usually better to read 'the newer papers first' when you work on literature survey. Often, there are students (beginners in research) who try to start from the older literature, thinking that they cannot understand the latest without understanding the earlier literatures. However, when one tries to start reading from the older literature, one would often be stuck by the arguments about the details of the experimental conditions or factors that cannot be taken into account in the simulations, or in other discussions that have been largely resolved and agreed upon later. It is usually advisable to read the introduction to the most recent literature to comprehend the current state of the art in the best possible way.

Tea Break

Why simulation collaboration cannot work well due to shortage of manpower: Researchers who make a living in a field really do not want their field to be invaded as a true feeling. Methodologists wish their methods would become more popular, because it would increase the demand for their activities and increase their value. If the method becomes too popular and more people are able to use it, the territory will in turn be invaded and its value will decrease. One would gladly cooperate with others if one's collaboration would lead to more coauthored projects, but if the collaborating partners will leave and become independent one day after the technology transfer, it would be like adding one more competitor by teaching them one's own tips. The honest answer is that one wants to cooperate, but does not want a complete transfer of one's technology. This mentality leads to a situation where the core part of a simulation is handle within the theory group's side, which eventually leads to a situation where the project is not able to move forward due to a lack of manpower. It was lucky enough for the author's group being mainly engaged in quantum Monte Carlo electronic structure calculations, not DFT itself as 'main occupation'. We use them use them in practice and are involved in their basic theory side, as such we were able to transfer the technology to practitioners without any problem of 'invasion of our territory', getting a good positioning. This has made it possible to collaborate without the problem of manpower shortage, as the collaborators on the experimental/industry side are able to perform the calculations on their own.

8.1.4   How to Perform Tracing Studies

Once you have found preceding simulation papers on the system you are interested in and have a simulation expert collaborating with you, you can proceed with the tracing calculations under the help of their advice. As we have noted in Sect. 1.3.3, beginners tend to start with their own complex and realistic target systems (surfaces, defects, etc.). However, even if you have calculated on a novel and seasonal system that has no reference for calibration, there is no way to verify whether your calculations are correct or not, and no way to develop insight into what has changed from the standard reference. This will eventually waste time and produce no results. It is strongly advised to find and start with an appropriate reference system which is closer to your final target (e.g., pristine bulk as the mother compound for the target surface or defect systems), for which reference results are available (e.g., lattice constants, bulk modulus, bandgap etc.). The first thing you are recommended to perform is work on the same calculations as the preceding papers report and see whether the same results are surely reproduced to check your computational conditions (this is a typical example to prevent from 'haste makes waste'). This type of the calculations is called **tracing calculation**.

Most of your practical interests would be in the surface/interface or problems involving defects or solid solutions. However, especially for beginners, one should rather start with pristine bulk, instead of your final surface geometry with such intention like 'I am studying surface systems'. Readers who have read this book through to this part would be able to understand the key specs reported in preceding papers (exchange correlation functional, pseudopotentials, basis functions, k-mesh, E_{cut}, etc.). Such readers are therefore able to perform tracing calculations under some assists by simulation experts in your collaboration. In the tracing calculations, you have to check whether you can reproduce band dispersions, density of states, lattice constants, cohesive energy etc., as described in Sect. 3.7.

After the reliability of your calculation conditions has been assured through the tracing calculations, one can proceed to what you want to describe (e.g., by gradually substituting elements on pristine bulk, or by cutting out your surface). This process should be done as if 'walking with knocking on stone bridges' as one does in programming development.[3] The standard format of analysis in electronic structure calculations would be to start with a reference system where the reliable calibration is assured (i.e., the past reported results are reproduced properly) and then to discuss what is changed such as 'the conduction band has shifted to lower in energy compared with that of pristine system', 'impurity levels have appeared', or 'surface levels have buried inside the valence band'. Therefore, it is never a detour to start with the calibration of a bit boring pristine systems, but it is the research itself that leads steadily to the publishable research.

---

[3] A typical example of this step can be found in the literature [1].

8.2   Tips for Successful Collaboration

8.2.1   Standardizing any Format

One of the crucial tips to successful simulation collaboration, especially remote collaboration, is to avoid wasting energy on troublesome technical details of ICT.[4] The main reason why collaboration stalls is that morale gets lowered due to this exhaustion occurring at unessential technical points. To avoid this, a tip the author's group has taken is to force all collaborators to use the same terminal environment, the same software, and the same operation techniques. Through out of this book, you could find several such 'forcing' instructions going too details (e.g., even the fingering of shortcuts), but these are manifestations of the author's policy as a preventive solution, which has been established upon the past unsuccessful cases with stalls in collaborations.

When conducting remote instruction under different terminal environments (Mac/Linux and Windows), one often hears from collaborators that they do not know how to set up Internet or printer connections, or that they do not know how to operate plotters which is different from each other. These are, however, matters that is not the main issue of the collaborative research, as such one wants to tell them 'figure it out and solve it on their own'. However, since beginners are also beginners in ICT technology, they are unable to move forward hindered by these stumbling blocks. As a result, the communication that used to be once every two weeks gradually becomes every month and hence every six months, which lowers morale to causes the collaboration to stall. It is important to be well aware of **important but not essential issues** which should be eliminated to prevent from stall in collaborations.

To avoid getting stuck in this kind of 'non-essential stuff', the author recommends to unify the PC environment and any manners of its operation to such extent where no freedom is allowed at all. If you allow collaborators to say, 'I'm used to this software, so I want to use this one', it will often end up costing you a lot in 'non-essential technical troubles' rather than the main contents in research.

For example, it is recommended to formulate rules within the collaborative group for presentation templates, fonts, use of colors, file naming conventions, etc., so that there is no unnecessary hesitation and efficiency is improved. Just as it is said that the reason Steve Jobs always wore a turtleneck sweater was to reduce the cost of thinking about the selection process, the positive aspect of a regulation is actually great to **reduce the cost in making decisions on your own** rather than the negative aspect due to the reduced freedom.[5]

---

[4] This is similar to winter mountain climbing. The most important rule is to avoid unnecessary physical exhaustion.

[5] Children with prodigious ability may feel comfortable when they are asked to draw a paint 'as you feel' on a given blank sheet of painting paper. The average children, however, would be confused by the infinite freedom of the blank sheet to write as one feels, but feels more comfortable with a restricted template to some extent.

Tea Break

When I mentioned that we have established rather strict rules and standardization of computer usage, an old professor who used to be a board member of directors once criticized me, saying, 'What you are doing is totally wrong because Science is an activity based on free ideas'. However, this kind of criticism is a typical failure to distinguish between what is essential and non-essential to research contents. In the field of experimental chemistry, the professor's specialty, it is a question of whether students should be allowed to do things in their own way or in different ways, such as arranging reagents in a cabinet, cleaning protocols of beakers, or the way of safely managing deleterious chemicals. Does forcing strict rules for such protocols hinder science develop with freedom atmosphere?

8.2.2 Pitfalls in Simulation Collaboration

The author's group has experienced several cooperative education on simulation, but not all have been successful. There have also been many collaborations that have stalled and vanished. Here, I would like to discuss what hidden factors may cause stalls based on our past experiences.

The first factor is the issue as explained in Sect. 1.1 that **(1) ab initio analysis is misunderstood as a whole simulation**, which would be the most common reason for the collaboration to fail at the beginning. For example, a story begins when an experimentalist working on complex surface reactions hears that an overseas competitor is using something called ab initio analysis, and he/she is very interested in doing the same. However, their motivation suddenly fades away when they notice that ab initio calculations can only give energies at zero temperature, or that they cannot directly deal with tiny concentration range that interest them. In order to avoid such tragedies as much as possible, we have provided an explanation on this point in advance in Sect. 1.1.

Secondly, **(2) the lack of young, hands-on staff** is another factor that is often seen as a stalling factor. Typically, such collaborations with industry domain that they just requesting the simulation analysis without any effort to work on simulations from their side are not likely to continue successfully. The aforementioned factors of insufficient manpower and lack of passion are the reasons for this. In addition, when 'people who have never actually tried running the simulation' are involved on a request basis, they cannot get a sense of the workload, which inevitably leads to excessive requests, which, together with the factor (1), leads to the impression that 'the simulation is not as useful as expected...' and lowers motivation.[6]

---

[6] This would be general issues for supervisors in a company organization. If a supervisor does not have practical experience, he or she will not be able to understand the workload as it really is, and there will inevitably be discrepancies in understanding the proper workload given for staff members.

Therefore, whenever we collaborate with experimental practitioners, we always ask to consider whether they can get younger members who are positive to learn the simulations performed by their own. By saying 'young' here, I do not mean the actual age, but rather the agressive mindset for learning new things without restricting oneself within what one majors so far. In fact, without such younger people who are actually doing the actual work, it ends up being nothing more than a conversation over a cup of tea about ideas. As such, whenever the author's group decides to collaborate, we first lend them a PC to share the work environment, and ask them to visit us to go through this tutorial course by their hands.

The most likely situation for success is where a group leader of a research group is so eager to have younger members acquire the simulation methods that he/she actually sends us his/her members to kick off the collaborative analysis. However, there are still many cases ending up with stalls. Surprisingly, most of the cases are caused by a factor that is far from the essence of the collaborating topic but purely technical stubbling that **(3) they cannot key-typing well at all**. In such cases, they open the lent PC only when they do a simulation (once several weeks), they cannot remember how to operate, running for picking up their hand-writing notes, and even when they know what to type, they don't use shortcuts, ending up with mess of mistyping accused by error messages... In the end, the member are still too lazy to use the PCs, and they are gradually losing touch with the PC terminals. The main reason why students in experimental group fail is **not because they find the theory too difficult**, but because they cannot do the touch-typing. In fact, students who have no difficulty with touch-typing are far less likely to stall.

In order to establish one's skill for typing and PC-use, a certain amount of time on training is required to be spent at the beginning. Taking the example of a SIM card in cell phone, its concept would be difficult to understand no matter how much you read the written explanations. However, these unfamiliar concepts are, after all, easy to get used to if you use them. Unless you use them, they will always be a mystery to you.[7] They need to take a certain amount of time to try their hands at it, but the students in the experimental group are just trying to hone their experimental skills in their main work and are not so motivated to spend time on matters that are not the main part of their work (I mentioned the same thing in my previous book). This is one of the key obstacles that can make the difference between a successful collaboration and a failed one.

---

[7] Programming skill is just like that. I sometimes heard that social scientists, who should be accustomed to using complex logic in their work, saying, 'I am a liberal arts major and have no idea about programming!', but the subject matter is actually understood even by junior high school students if they work on it with their hands. This is the reason why we strongly demand that 'young' people be the ones to intervene in the collaboration. Once a person has become a position in one's 'built castle' as one's past achievements, the pride on the castle hinders the person to learn new things, which the person tend to assume that one must not be able to understand, but even a junior high school student can understand it.

8.2.3 Pitfalls in Reporting and Consulting

Another major stall factor is **(4) the lack of frequent reporting, communication, and consultation**. After the long silence without any communication from collaborating students, I asked him what he was doing, then he replied, 'The calculations are going around, but the results obtained after each three day are always strange, and that's why I'm stuck'. Then we checked his out files, and found that the SCF calculation should not have taken such long CPU time like 3 days for such a small size of his system. Regardless of the three-day job limit of the supercomputer, it was in fact that every calculation had terminated abnormally immediately after it started running due to a typo-mistake input file.

Beginners understand the situation as 'simulations run successfully but the results are strange' just because the job actually started running and the out file surely came back (everything is as expected!). They are not even taught to check for the string, 'error termination', at the end of the out file. Or even if the energy value that comes out is positive, some think it is running normally because 'the value surely came out'. Even if there is an implicit understanding among theoreticians that the energy values in this field is usually negative, it may not be self-evident knowledge to beginners.

This can be easily avoided if one is able to frequently report progress and consult with specialists. However, it may not be possible with the standard mentality of Japanese students to casually consult with a superior professors introduced by their boss, whom one does not know well in a remote communication.

The author's group has experienced several remote international collaborations, but remote collaboration is much more difficult than one might imagine. When a group grows beyond a certain size, it generally becomes busy in hundling collaborations those are running simultaneously in parallel wise. If three weeks go by without any notice, the collaboration gets to be forgotten in such a busy atmosphere. However, when the collaboration is left to the younger students, such situation can easily occur and the collaboration is on the verge of disappearing. When we tried to find out the reason, we found that they had posed a question to their external professors, but were waiting for a response (maybe the external professors just forget it), or that the answer they received from the professors was a little off the mark, and they were too afraid to address it, and stuck with no idea how to write back in email. Younger students cannot boldly ask the professors, "I'm still waiting for your reply", or "That's a little off the mark and doesn't answer my question. What do you mean?". In the author's group, we first make our students learn how to compose such reminding/correcting emails in such wording that they don't feel impolite.

It is also not good practice at all to make no communication, saying 'Nothing to be reported because the calculations are running smoothly with no particular problem and progress'. It is very important to make the collaborators be aware of the existence of the collaboration with you by sending an email reporting 'There is nothing in particular, but things are running smoothly'. In fact, such report that 'progress is going well without any problems' is actually dangerous. There was a case in which we believed a student reporting that there were no problems, and when we inspected the achievements six months later, we found that the operation had been done completely wrong and had turned into a pile of garbage. Since the student was reasonably competent in understanding the theory and computer operations, we had left him to his own devices at ease, that turned out to be wrong in this case. Just reporting "it's going well" repeatedly, he did not consult with others much, being actually not very good at bringing up specifics for discussion with disclosing the contents of his work. What was worse, all the inputs and outputs of the calculations had been discarded, leaving only graphs based on incorrect calculations, making it impossible to trace back the data. Learning from the above case, **it is quite essential to use shared storage as possible so that all the collaborators can check any (input/output) results.** If an instructing collaborator can immediate access to the set of files containing the results of the calculations, he/she can easily find any mistakes using 'grep' [1] etc.

For the academia-basis collaborations, it is easy to adopt such a shared file server such as 'Dropbox' [3]. In the case of industry-academia collaboration, however, this kind of file sharing becomes quite difficult due to security concerns. In order not to stall collaboration, we must first secure how to devise and set up a file sharing system. This is another example of a 'not essential but important' matter mentioned earlier, repeatedly.

As mentioned above, simulation collaboration often stalls because of purely technical factors being not essential to the theory of electronic structure calculations. In this book, I have explained a little too much about the matters that are not essential such as 'keyboard shortcuts', 'the use of plotters' etc. This reflects our experience in realizing the above-mentioned 'not essential but very important' points to prevent from stall.

8.3 Tips to Train Practitioners

8.3.1 Computational Operations Are Not the Hard Issue to Non-experts

Throughout this book, we have emphasized the importance of operational technical matters such as typing and PC use. The command line operations appeared in this book gets to be rarely mentioned in general education these days[8]. In the author's previous book, I mention this misconception in terms of 'a misleading belief of the conservation of capabilities' [4], but Linux utilization is just the 'utilization technology', not requiring any special higher education. Actually in these days contrary, it is rarely taught even in Computer Science departments, but quite likely that only the students in Quantum Chemistry or Computational Physics would contact with it.

---

[8] When the author was an undergraduate, general education in engineering departments related to the use of computers surely included the use of UNIX, with chapters such as 'Using Electronic Mail', where emails were handled by UNIX command operations at that time.

Let us assume such a reader who wants to introduce simulation as a group organizer, rather than a young practitioner who handles the operations himself. He may be wondering how to secure a human power with advanced UNIX operation knowledge. I note, however, that in my research group, the simulation operation itself is handled by the research assistants who are actually part-time assistants. We have had three assistants, all non-science majors who graduated from local universities, such as English literature or junior college, and who had only Windows/MS-Office skills as staff in the common office taking care of clerical issues for all faculties members before joining the author's group. They have mastered Linux commands within six months after joining our group, and now they can handle not only 'grep/awk/sed' [2] but also gnuplot, TeX-typesetting [5], and even script work without any problems. These research assistants are in charge of the data operations that resulted in the author's original paper [6] in Phys.Rev.Lett.

8.3.2 Aspects of Work Sharing

Experts armed themselves with a degree in simulation science are needed only when one has to compose the discussion part for the obtained results, and to plan and design the whole structure of a simulation study. Once the overal plan has been established, individual operations do not require any particular 'science degree' background, as emphasized above. In the 'education mission' for Ph.D., it is important to make one be able to complete a series of tasks on their own, not only talking about the idea and concept on the background literature survey, but also everything required, such as, how to formulate the problem, preparation of tools to perform the formulation, handling the operation to get data, data analysis, developing possible discussions on the data, scenario planning based on the positioning of the research etc. However, in the practical/industrial projects, there is no reason to refuse to divide all labor over the working staff in order to improve efficiency.[9]

It is important to realize that the operational part is not something that requires scientific training in higher education, but is something that can be mastered simply through habituation. Once this is realized, project managers can focus their search on people who can touch-type, who do not limit their own abilities within what they can do so far, and who can perform the job precisely with less mistakes. By assigning such people, the simulation practice will start running without any problems. It is surprisingly possible to find such personnel for simulation operations among part-time assistants who can do excellent clerical work.[10]

---

[9] In fact, having a non-specialist in charge of operations has the advantage of maintaining objectivity without any preconceived notions.

[10] On the other hand, some of the post-doctoral researchers are often the case that they are unable to adapt such works because of their obsession/pride with the field they define as their specialty, or because they tend to stick their self-imposed sense of weakness for computation as their prideful identity.

The same story applies to server administration in a research group. In my previous book [4], I mentioned that rather than waiting for students who are interested in and capable of server administration to appear, we have been rather successful in training all the normal students who are not so interested in server administration at the beginnig to become server administrators. Minimum required Linux operations are explained in the author's previous book as well. There are students from experimental groups who can learn and master the Linux operation on their own without a tutor by reading the book.

References

1. "Bandgap reduction of photocatalytic TiO_2 nanotube by Cu doping", S.K. Gharaei, M. Abbas-nejad, R. Maezono, Sci. Rep. 8, 14192 (2018). https://doi.org/10.1038/s41598-018-32130-w
2. "Linux Command Line and Shell Scripting Bible", Richard Blum, Christine Bresnahan, Wiley (2021/1/13) ISBN-13:978-1119700913
3. https://www.dropbox.com/official-site (URL confirmed on 2022.11)
4. "Jisaku PC cluster tyou-nyuumon", Ryo Maezono, Morikita Publishing (2017/12/14) ISBN-13:978-4627818217
5. "Guide to LaTeX (Tools and Techniques for Computer Typesetting)", Patrick W. Kopka, Helmut Daly, Addison-Wesley Professional (2003/11/25) ISBN-13:978-0321173850
6. "Excitons and biexcitons in symmetric electron-hole bilayers", R. Maezono, Pablo Lopez Rios, T. Ogawa, and R.J. Needs, Phys. Rev. Lett. 110, 216407:1-5 (2013). https://doi.org/10.1103/PhysRevLett.110.216407

Appendix A: A Short Course of Linux Command Operation

9

Abstract

This appendix chapter is introduced from the main text at Sect. 2.5.2. This appendix provides a short course of Linux command operations, which is the biggest obstacle for beginners. At the earliest user level, UNIX and LINUX can be used to refer the same [1]. Since UNIX has been used to build huge systems, it have been refined over its long history, offering a variety of tools that enable 'mass processing/automatic processing/high-speed processing'.

9.1 Getting Familiar with Directory Structure

9.1.1 Directory Structure Instead of Folders

The location[1] of the file "setupMaezono.tar.gz" as downloaded in Sect. 2.4.2 is described as '/home/student/work/setupMaezono.tar.gz'. This is just like to specify the location of the author as 'japan/ishikawa/nomi/jaist/information/maezono'.[2] This is a way of the most formally defining the location of the file, and is called **absolute path** specification because it absolutely defines the path to the location. In other word, it can be 'the most official name' of a file.

The format, 'xxx/yyy' can be interpreted as 'yyy' registered on the 'directory booklet'. This is the reason why 'yyy' is described as 'yyy under the xxx-

---

The original version of this chapter was revised: Belated corrections have been updated. The correction to this chapter is available at https://doi.org/10.1007/978-981-99-0919-3_14.

[1] The usage of the commands is almost the same. The differences are at the professional level in how to implement this command system. In a metaphorical way taking car driving for example, you don't need to worry about whether your training car is a Toyota or a Honda at the earliest level. Once you become proficient, then you can distinguish the differences between Toyota and Honda.

[2] The author is currently working as a faculty in School of Information Science of JAIST (Japan Institute of Science and Technology) located at Nomi(city), Ishikawa(prefecture), Japan.

R. Maezono, *Ab initio Calculation Tutorial*,
https://doi.org/10.1007/978-981-99-0919-3_9

directory'. For the full-path specification like 'jaist/information/maezono', it is almost likely that 'maezono is obviously at jaist', so it is convenient to abbreviate as '~/information/maezono'. For '/home/student/work/setupMaezono.tar.gz', it is usually implied that the location is always under '/home/student' directory (home directory). As such, it is usually denoted as '~/work/setupMaezono.tar.gz'.

Start up 'terminal' app, and execute

```
% cd
% pwd
```

You will get '/home/student' on the terminal as the 'responded value' to your input (pwd) by the system. Next,

```
% ls
```

leads to the response showing 'Desktop/ Documednts/ ...' on the terminal.

The command "pwd" is an abbreviation for "present working ddirectory", which returns the location of the directory you are currently "descending" (the current directory location). Most Unix commands are abbreviations for meaningful words like this, and these are called **mnemonics**. That's important to get into the habit of remembering mnemonics, and then Unix commands will become much more familiar.

The command "ls" is a mnemonic for list, which displays the files and folders directly under the current directory. Since "directory" means 'directory booklet', "ls (list)" would originally corresponds to 'showing the list on the page in the booklet'. Rather than that, however, it is better to take the imagination that 'ls' is used to see the view from the current directory you are descending.[3] That's all for the explanation. Then, type "exit" to exit the terminal.

Let me give the general caution in advance for learning Linux commands. When I teach those students being not familiar with PC, I often find that they tend to write down every commands in their notebooks. They are quite likely to face trouble, whinging that "I typed it exactly as you taught, but I got an error...". The error occurs because they execute it for 'file01' with specifying 'file_example_name' that I taught as example. Their executing commands are sometimes lacking a space between [command] and [argument] because it is based on the hand-writting on their notebook (space is not explicitly shown there). Do not treat the commands as incantations (typing without thinking). Consider the meaning and build a structure of your command along your logical consideration, remembering the fundamental rule as '[command] [-option] [argument]'.

The following is a typical example which confused the author when I was a beginner (no need to execute here):

---

[3] Taking the metaphoric of the booklet, 'directory01/directory02' becomes 'booklet02 in the list on the booklet01', that's weird.

```
% sed 's/\t/\//g' temp1 > temp2
```

I could not recognize any meaningful and logical structure in the sequence of '\' and '/'. The above command can equivalently be written as (no need to execute here)

```
% sed 's#\t#\/#g' temp1 > temp2
```

Comparing these equivalent expressions, you notice that '/' (='#') plays the separater while '\t' and '\/' are the characters [special characters; '\t' is the 'tab' (usually invisible), and '\/' is the '/' as a character (not playing as separater)]. If a beginner operates with the attitude like "Just type what it says. This is just a task", one will inevitably end up making mistakes like, "The number of '/' is not correct". Not like that, but typing with considering "'\t' and '\/' should be separated by '/'" is the desirable way for the operation.

9.1.2 Moving Between Directories

Start the terminal again and execute 'pwd'. You will see '/home/student', which corresponds to the '~/' location explained above. Here we notice that, when you start up a new terminal, you will descend to the '~/' location. This is the reason for '~/' to be called 'home directory'. Next, execute 'ls' to look at the 'view from the descending position'; you will see items such as 'work', 'Documents' etc. Here one notices that they are indicated by a '/' at the end, which means "This is not a file, but a directory with a substructure inside it.

Next, execute

```
% cd work
% pwd
```

Be careful about a space between 'cd' and 'work'. This time, 'pwd' responds as '/home/student/work'. Let's try understanding this: By "cd work", you "moved" your position under the 'work' directory, which is one of the further directories visible from your home directory. As the result of this move, the 'pwd' (present working directory) is modified as '/home/student/work' = '~/work'. 'cd' is a mnemonic for "change ddirectory". The **argument** to 'cd' is used to specify the name of the directory you want to move to.

Moved under 'work' directory by the above operation, execute 'ls' to see a list of files and directories as a view from where you are standing. You will find "setup-Maezono/" with a slash at the end, indicating that this is not a file but a directory with a further substructure. Then, execute

```
% cd setupMaezono
% ls
```

to go down into 'setupMaezono' directory. By the 'ls' command, you can check the 'scenery' viewed from your 'standing point', finding 'us' directory. Going further down into 'us', you can find 'california' directory, below which you can further two directory named 'caltech' and 'stanford'.

Now, let's get into 'caltech'. You will find a file, 'feynman' (no slash at the end).[4] Execute 'pwd' here, and you will get reply[5]

```
/home/student/work/setupMaezono/us/california/caltech
```

Executing 'ls' results 'feynman' as the response. The file is formally specified as '.../us/california/caltech/feynman' in the full-path manner.

Next, execute

```
% cd
% pwd
```

The response will be 'home/student', from which you can understand that 'cd' takes you back to your home directory when it is executed omitting the argument. In Sect. 2.5.2 of the main text, I mentioned that beginners can easily get lost in the directory structure, but even if you do get lost, you can come back to your home directory by using the 'cd with no arguments' command. Now you can wander around the directory without worry.

Then, go down to '~/work/setupMaezono/' again, this time to the directory where "us/california/stanford/bloch"[6] is located. Now, if you are asked to execute "cd" to go back to your home directory again, and asked then to go down to "us/california/stanford/" again, you may feel a little bored or exhausted especially for a beginner who is not accustomed to typing. It is really tiring to type "california" or "stanford" every time without making any mistakes. So, I am going to explain **the most important** thing to be leant now. Anyway, please go back to your home directory by executing 'cd'. Then, type 'cd De' and try your best to press the tab-key with the little finger of your left hand. It should automatically be completed as "cd Desktop". This is called **tab completion**. Hit enter-key and then type "cd" again to return to your home directory.

In the above, you typed up to "cd De" to get tab completion, but now try typing up to "cd D" to get tab completion. You will notice that the completion is not successful. Hit the tab twice, and you will see "Directories starting with D (Documents and Downloads)" proposed. Tab completion narrows down the candidates to "up to the string you typed" and completes when it is the only one. Until you get to the only

---

[4] R. P. Feynman is a Nobel laureate in Physics and was a professor at the California Institute of Technology (CALTECH) in California, USA.

[5] The backslash '\' is used to show a too long string to fit on a line (see Acronym).

[6] F. Bloch is a Nobel laureate in Physics and was a professor at Stanford University.

one, just keep hitting **tab** to narrow down the list of candidates.[7] Then, go back to your home directory again and use tab completion to go down under the 'work' directory.

When you come under "work", execute "ls" to check the contents of the folder. Next, "cd s" for a tab completion and get into "setupMaezono" and check the contents with "ls". Then use "cd u" for a tab completion and get into "us". 'cd' to get back to your home directory and repeat the same thing several times to learn tab completion **with your fingers**. When you are sufficiently proficient with tab completion, go back down to "us/california/caltech".

> **Important**

Tab completion is an extremely important skill for fast typing without mistakes. If you are used to typing on a word processor, you may be familiar with "tab manipulation with the little finger of the left hand" already. In recent years, however, younger students who cannot type on a keyboard has increased due to the spread of smart phones. When they joined our summer school, they are repeatedly pestered as 'use tab!', but the young students are so adaptable that they say, 'I don't want to use the mouse anymore' after three days of summer school.

Then, this time, let me ask you to move from "Caltech with Feynman" to "Stanford with Bloch". If you only use what I taught you, you would go back to your home directory by using 'cd without argument' as a 'talisman', and then go down under work again. This is, however, very inefficient, just like going out of US and entering the US border again whenever you want to go from Caltech to Stanford. I will teach how to avoid soon, but for the time being, move to Stanford by this inefficient way.

Then, let's think of a better way to get from Caltech to Stanford. The usual way would be to go from Caltech (california/caltech) back to California, which is one level higher, and from there down to Stanford (california/stanford). The command to express "go one level higher" is 'cd ..' (note the space after 'cd'). If you use '..' (two consecutive dots) as the argument it means 'one level up' in the directory structure. Remember "." (one dot) means "the current directory". ".." means "the directory one level higher (called parent directory)". Then execute 'cd ..' and 'pwd' to see where you are. Try go back and forth between feynman and bloch until your finger remember how it works. After that, get back to your home directory.

Then, from the home directory, execute

```
% cd work/setupMaezono/us/california/caltech
```

Be sure that you are using tab completion. Type 'ls' to confirm that you are now at Feynman's place at CALTECH. As this example shows, you can specify your

---

[7] We do the same thing when we type on a mobile phone. For example, if you want to type 'Hello', you will type 'Hel' and then you can pick up the proposed candidate 'Hello'.

destination (if you know that) and jump directly to it at once, avoiding to go down the directory step by step. Taking this way, you can start from Caltech to Stanford by 'cd ../stanford'. Or you can go there from anywhere (including Caltech) by

```
% cd ~/work/setupMaezono/us/california/stanford
```

This is just as two different options to specify your destiny asking for a cab by saying "Stanford around that corner" or asking the driver to input the address on the car navigation system.

> **Important**

Whenever you are instructed to move to somewhere, make it a habit to use 'pwd' or 'ls' as

```
% pwd
    /home/...[Omitted].../us/california/caltech
% ls
    feynman
```

to confirm whether you have moved to the intended directory and the status of the working directory before you execute further task.

9.1.3 File Operations

Go to the stanford directory and look at the contents of the file "bloch". As follows,[8] 'cat [file]' can do that ('cat' is a mnemonic for 'catenate').

```
% pwd
    /home/student/work/setupMaezono/us/california/stanford
% ls
    bloch
% cat bloch
    ...(omitted)
    192.168.0.250    i11server250
    192.168.0.251    i11server251
    ...(omitted)
```

---

[8] The backslash '\' is used to show a too long string to fit on a line (see Acronym).

> **Important**

By starting with 'pwd' as above, you can accurately convey the following information: 'In this location, there is this file (displayed as 'ls'). Looking at the contents
of the file, ...'. This manner is quite useful when you consult with a skilled person
with your troubles, or you instruct others to perform various operations. The manner
enables accurate communications with least words by showing a series of **appropriate command input/output**. This way, you can avoid problems occurring in most
cases with beginners, i.e., the errors due to the commands executed at improper
directory locations ending up with 'target file/directory not found'.

Now, what was written at the beginning of the file 'bloch'? Using the mouse
to scroll up the file, you will find that it is very long. When displayed by 'cat',
everything is displayed instantly, and unless you have very good eyesight, you will
miss the beginning. Instead of 'cat', try

```
% more bloch
```

Then the display will start from the beginning, and you can use the return key to
move forward one line at a time, or the spacebar to move forward several lines at a
time. The display will end when you reach the end of the file.

By the way, if this file is a long file with contents that you don't want others to
see, and suddenly a person appears behind you when you browse it by 'more', what
should you do? You want to stop displaying the file, but no matter how hard you
press the spacebar, the file is too long and the embarrassing content continues to be
displayed. In this case, you can use '**Ctrl+C**' to **force quit**.

Perhaps due to the influence of Windows-OS, there are many beginners who
mistakenly use 'Ctrl+Z' intended to force quit. So, we need to clarify the difference
between the two. Again, open the file as "more bloch" and try "Ctrl+Z". The display
will disappear as it did with "Ctrl+C", but you should see "[1]+ Stopped...". Then try
typing "fg" (a mnemonic for underlinefore underlineground) and then you will see
that the display of the file 'bloch' gets back to the operation again. Here you notice
that "Ctrl+Z" does not work to force quit, but puts the process on hold (**pending**), and
"fg" restores the pending process back. Even if the process is pending, it disappears
from the screen, so beginners often misunderstand this as a forced termination. As
a result, they abuse "Ctrl+Z" to produce a pile of pending tasks, resulting in a large
number of pending processes accumulating and **overloading the system**. This type
of trouble can be found frequently, so be careful not to mix up "Ctrl+Z"(pending)
and "Ctrl+C"(force quit).

Again, go to the stanford directory and execute

```
% cp bloch knuth
```

By 'ls', confirm if a new file 'knuth'[9] has been created. '**cp [file01] [file02]**' means 'copy file01 to a file named file02' taking a form of 'command [file01] [file02]'. By the above, the contents of the 'knuth' should be the same as those of 'bloch', so please confirm it with 'more'.

After confirming the above, execute the following under the same 'stanford'

```
% mv knuth knuce
```

Execute 'ls' to confirm that 'knuth' has been renamed to 'knuce'. Execute 'more knuce' to confirm the contents unchanged. The command 'mv' (mo_ve) takes two arguments, meaning 'rename the file of the first argument to the file of the second argument'.[10] Note that many beginners tend to confuse it with cd (=change directory) and type 'mv' when they should type 'cd' (maybe due to the image from the word "move"). Then, in the same directory, execute

```
% rm knuce
```

Confirm that the 'knuce' has been removed (rm is a mnemonic for re_move).

The commands including 'cp' do not limit the argument to a file under the current working directory. You can execute

```
% pwd
    /home/student/work/setupMaezono/us/california/stanford
% cp bloch ../caltech/pauling
% ls ../caltech/
% more ../caltech/pauling
```

to operate the distant files[11] (Be sure you are using tab completion). The 'more' executed above is equivalent to

```
% more ~/work/setupMaezono/us/california/caltech/pauling
```

Note the difference that the specification, '../caltech/pauling', works only when your current working directory is directly under the same 'california' directory, while the latter specification works no matter what directory you are in. In this context, the former is called "**relative path specification**" while the latter "**absolute path specification**".

Next, getting into the 'california' directory, make sure you can see the 'stanford' and 'caltech' directories there, and then execute

---

[9] D. E. Knuth is famous mathematician at Stanford University, also known well as the developer of TeX.

[10] It's okay to write this way now, right?

[11] L. C. Pauling was a Nobel laureate in Chemistry, and was a Professor at CALTECH.

```
% mkdir berkeley
```

By using 'ls', confirm if the 'berkeley'[12] directory has been generated. The command "mkdir" is a mnemonic for "<u>m</u>ak<u>e</u> <u>dir</u>ectory", which creates a new directory with the name specified as the first argument. Then, execute below at the same location,

```
% cp stanford/bloch berkeley/
```

Be careful about spaces between arguments, and be sure to use tab completion (it automatically puts '/' at the end). By executing

```
% ls berkeley/
```

confirm if the file 'bloch' has been copied under the directory. Confirm further if the contents in the files are the same by using 'more'.

Next, let's try to delete the newly created 'berkeley'. The delete command was 'rm', but even "rm berkeley" would not delete it. Since berkley is not a file, but a directory, we need to add the **option "-r"** to remove it[13] :

```
% rm -r berkeley
```

Next, make sure that you are directly under the 'california', and then use "cd .." to move up a level and execute

```
% cp -r california newyork
```

In the above command, the 'california' directory is entirely copied to the 'newyork' directory. Again, the option "-r" is used, which means "make the target a directory".

Next, make sure that you are directly under the '~/work/setupMaezono/us' directory and that you can see the 'california' and 'newyork' directories, then carefully and correctly type and execute the following:

```
% cp california/caltech/feynman .
```

Be careful about the final dot '.' and the space before it. Remember that the second argument "." means 'the current directory'. So the above command execute to copy 'california/caltech/feynman' to 'here'. After confirmed by 'ls' that the 'feynman' has been created in the current directory, delete the file using 'rm' and delete the 'newyork' directory as well.

---

[12] California State University, Berkeley exists in California.

[13] Let's look up and understand why the mnemonic "-r" is used for the option to specify a directory.

9.1.4   Editting Files

Type 'cd' to return back to your home directory and execute

```
% emacs -nw work/setupMaezono/us/california/caltech/feynman
```

The command 'emacs' starts an editor for editing a file specified as its argument. In the above, it opens the file 'feynman'. The "-nw" option corresponds to a mnemonic, "no window", which means to open a file in the terminal deck instead of opening a new window.

Once the file is open, the first thing you need to learn is how to close it. Type "Ctrl+X" and then "Ctrl+C" to close the editor. It can be equivalent to 'Ctrl+XC', but it is safer for beginners to make it separately as "Ctrl+X" and "Ctrl+C", because you will learn such operation as "'Ctrl+X', **release it**, and then type only 'I'" afterward, which will make you confused whether 'XI' or 'X (release) I'.[14]

In this book, we will use the 'emacs' editor, but there is another typical editor called '**vi**' [2]. Just as PC users are divided into Mac/Windows, the user's preferences for editors are divided into 'emacs' or 'vi', sometimes discussed passionately. If you want to become a full-fledged server operator, it is recommended that you become proficient in the 'vi' editor at some point. 'vi' is compact enough not to burden your system, so it is always included natively in every system, while 'emacs' is a relatively heavy application, as you can see from the fact that it was installed from the network. This is one of the reasons why it is avoided from professional viewpoint putting importance on "being able to use it in any environment". However, 'vi' retains the usability of older editors, i.e., the command system like 'erase the Yth character of the Xth line' before the intuitive usability with arrow keys to move cursor position for input. So even erasing a single character can be too difficult for beginners. This is the reason why we take 'emacs', which is highly intuitive for beginners.

Then, we want to open the 'feynman' again using "emacs -nw ...". Here, type up-arrow key just once, and you find that **the previous command executed appears again**. Typing it several times shows the commands you have done so far one after another. If going too far, you can use the down-arrow key to go back. By using that way, open the 'feynman' again, doing nothing, and close it with "Ctrl+X" and "Ctrl+C".

Then, try executing

```
% history
```

---

[14] As you become more proficient, your fingers will "learn" these operations instead of your brain. This is why, when teaching a beginner, even an experienced user often asks his/her fingers, "How did I do that?". It is similar to the way a person who is proficient in 'abacus' calculates by moving his/her fingers.

It responds with the list of previously executed commands with the number at the head of each line. Identify the line number of the command you did previously

```
% emacs -nw work/(...omitted...)/caltech/feynman
```

Let's say the number is 63, then execute

```
% !63
```

You will find that it executes 'emacs -nw ...(...omitted...)/feynman' to open the file. In this way, you can use the 'history' command to check what you have "done so far", and then re-run past commands by number as identified.

Though we have just tried 'opened and closed' the file so far, now that, let's try to **edit** it. When you open the file in emacs, you will see the string "-UUU:----F1" in the white strip at the bottom left. Next, place the cursor at the beginning of the file and press the return key to create a blank line at the beginning of the file. Then write your name (e.g., student) on that line. If you look at the string in the white strip, you will see that it has changed to "-UUU:**--F1" with an asterisk (*). This asterisk indicates that the file has been modified, but not yet saved. To save your changes, execute the "save operation", **"Ctrl+X"** → **"Ctrl+S"**. When you do this, the white strip will indicate "Wrote..." and you will see that the asterisk "**" has disappeared. This means that the changes have been saved. Then, close the file and reopen it to confirm that your name is shown there, corresponding to that the previous modification has been saved properly.

Then, erase your name to restore the file to its initial form, and save it again with "Ctrl+X" → "Ctrl+S". Next, write your name on the top line again in the same way as you did. From the cursor at the end of the first line, try **"Ctrl+A"**. The cursor will move to the beginning of the line. Then **"Ctrl+E"** will move the cursor to the end of the line. Once your fingers remember these, "Ctrl+A" again to move to the beginning of the line and then try 'Ctrl+D'. Once you get used to using "**Ctrl+D**", you will no longer have the habit of typing delete key repeatedly to get back to the end and delete the target part, leading to your work being more efficient. Writing your name again, and use "**Ctrl+A**" to return to the beginning of the line, and then try 'Ctrl+K'. This is a shortcut for "erasing everything after the cursor in a line". If you press "Ctrl+K" again, the blank line itself will disappear.

Next, let's add two or more lines from the top line, writing your name and your friend's name as well. Get into the habit of saving the modification using "Ctrl+X" → "Ctrl+S" whenever it occurs. Next, move the cursor to the beginning of the file. Then press Esc-key once, release it, and then press "shift+>". (since "typing >" implicitly includes "pressing shift", we omit "shift+" hereafter, describing this operation as "**Esc+>**"). The cursor will jump to the end of the file. For beginners, it would be difficult at the beginning to follow such key combinations until your fingers get used to do it. During the practice, if you make some mistake and cannot get back to editing, try **"Ctrl+G"** several times to get back to the previous status (this is also an important 'amulet' to remember). Once your fingers have learned the above "**Esc+>**", try pressing "Esc+<" with the cursor at the end of the file. The cursor will then jump to the beginning of the file. Repeat the exercise a little until your fingers learn to move to the [beginning or end] of the [file or line].

Finally, the Undo command is "(Ctrl)+(shift)+(-)". Try it out on the document you are editing as appropriate. These above are the minimum **shortcuts** for the emacs editor. A search for "emacs shortcuts" on Google will give you more detailed information. Even if the shortcuts are tedious at first, it's important to make your effort to learn them. A few seconds/minutes of the effort surely awarded by saving the years in efficiency.[15]

9.2   Use It More Conveniently

9.2.1   Using Alias

Linux beginners may start getting the impression that 'it is troublesome to remember commands' (although tab completion is useful), but even proficient users do not remember and use all the huge number of commands. In fact, we can 'register' combinations of frequently appearing commands and use them conveniently to your own specifications.

For example, for "emacs -nw" with the option "-nw", remembering and typing this every time is a burden because it increases the amount of typing. This is a simplest example with only one option, but in general, more complicate combinations of options appears those are impossible to remember in practice. Here, let's try

```
% alias enw="emacs -nw"
```

and confirm that executing "enw" actually works as "emacs -nw". This kind of "redefinition of commands" is called "aliasing". As another example, such a sequence with ['cd ..' to go up one level and check the contents by 'ls'] is quite frequently occurring. Then, aliasing as

```
% alias cup="cd .. ; ls"
```

can work as the sequence by typing just a command 'cup'. Here you notice that '[command1];[command2]' can be a one-line execution for a sequence of two commands connected with colon ';'.[16]

---

[15] It's the same as training others staff at work. Since it takes a time, one tends to avoid it, thinking that it is faster to do it by oneself than teach them. However, even if it take a bit time at the beginning, he/she will become another horsepower in a few days, and in the long run, it will lead to great work efficiency.

[16] The examples explained here, 'enw' and 'cup', are not actually used. These are aliased as 'mnw' and 'c', respectively, by the setting in Sect. 2.4.3, and used in the tutorials starting from Sect. 3.

Now, exit the terminal by executing "exit" (we describe it "exiting from the terminal"). Then, start the terminal again to see whether "enw" and "cup" are working. You will get an error whinging "No such command found". Even if you have set up aliases, the settings are forgotten whenever you exit the terminal, and you have to reset the aliases every time. However, it would be very impractical to reset a large number of aliases by typing each time you log in. In actual operation, you prepare a file on which a series of aliases are 'registered', which is located somewhere, and let the system 'activate' the file automatically to make aliases ready to use whenever you log in.

Now browse the file '~/work/setupMaezono/example_alias' by using 'more' (be sure to use tab completion!). In addition to 'enw' and 'cup', you can see that there are several other aliases registered on the file. For such a alias file, executing '**source** [alias file name]' makes it activated. For the present case,

```
% source ~/work/setupMaezono/example_alias
```

activates the aliases on that file. After that, you can confirm that 'enw' and 'cup' can be used now. Nevertheless, it is still tedious to remember where the alias file is located, and execute 'source' command whenever you log in. Usually, there is a convention for the location of the alias file, and the way to make it automatic to execute 'source' as I explain below.

Go back to your home directory and execute 'ls' first to get the response. Then try

```
% ls -a
```

with the option '-a'. Comparing with the response without the option, you find that more files or directories are displayed those names starting with dot (.) such as '.cache' etc. These files or directories with a dot (.) at the head of each name are called **invisible file**. Beside the 'normal' files (document files user has created explicitly), there are many 'system-related' files on which system setting specifications are defined. These system files are treated as invisible by putting a dot (.) on each filename so that they cannot be displayed by default without special options.[17]

The alias file is therefore treated as an invisible file, and usually it is placed as "~/**.alias**". As such, execute[18]

```
% cd
% cp ~/work/setupMaezono/example_alias .alias
```

They are already usable even now. In order to explain how aliases work, we have to use the different names, 'enw' and 'cup', which are not set in the aliases.

[17] In our daily life, we can see e.g., a refrigerator or a chair in your room, but there also exists molecules of smelly gas, or a person as a ghost etc., those are inconvenient when displayed.

[18] Since the 'source' command is executed with the same argument right before, it would be great if you could go back with the up-arrow key, press 'Ctrl+A' followed by 'Ctrl+D' to delete the word 'source', and then type 'cp' followed by 'Ctrl+E' to move to the end of the line, and add the '.alias'.

Execute 'ls -a' again to make sure that the '.alias' is created on the home directory, and check the contents of the '.alias' using 'more' to make sure that it has been copied correctly. After that, execute 'exit' and close the terminal.

Starting up the terminal again, first make sure that 'enw' and 'cup' are not working. Then, execute

```
% source ~/.alias
```

and make sure that the alias is activated so that 'enw' and 'cup' can work now. In this course, we will use this alias file.[19] If there is a new alias that you want to register, you can add it by following the existing format in the '.alias'. You can then raise your alias file as you want.

Tea Break

The alias file can be personalized as a matter of course, but it's better to make it shared among the group member or research community. It greatly accelerate collaborative work. In my case, the alias shared in my group was based on that originally composed by my ex-colleague (Dr. M.D. Towler) when I was a post-doc member in some institute (Cavendish Lab.). It is quite moving when I find the same alias in different group far abroad, implying the same bloodline spread by researchers when they were young postdocs jumping over the world.

9.2.2   Pipe and Redirect

One of the useful features of Linux is the ability to feed the output of one command to another command as input. This is a fascinating feature that should be on board in "Noah's ark".

Go to your home directory and execute "du -h .cache/" here. Responding texts will flow on the screen for a while. and you can use this command as The command, "du [directory name]", is used to display the capacity of files under the specified directory ("-h" is an option to display the capacity in kilos or megs).[20] Because the size of the invisible directory ".cache" are considerably large, the responding texts scrolls so quickly that it runs away before one can finish reading it. To read the first part, you have to use the mouse to scroll it back. Now, try the following with a vertical bar.

---

[19] This seems to require that every time you start a terminal, you have to make sure to be at your home directory and execute "source .alias", being tiresome. There is a more convenient way to make it automatic, as explained in Sect. 9.2.3.

[20] 'du' is a command used for system administration. Note that there is no intention to teach the detailed function of 'du' to beginners in particular, but just as an easy example for using pipe.

```
% du -h .cache/ | more
```

The display will stop after one page of screen is displayed. After that, you can use the "more" operation to display one line more by enter-key and one page more for space-key.

The vertical bar connecting two commands as, '[first half command]|[second half command]', is a convenient way of using Linux, working as 'passing the output of the first half to the second half as its input'. It is described as 'the first is piped to the second'. In the current example, lengthy one-shot flow from "du -h" is piped into 'more', which can display the contents step by step manipulated by enter- or space-keys.

The 'redirect' function provides a similar level of usefulness (worth to be loaded on "Noah's Ark") as explained below. Making sure to use tab completion, execute

```
% cd ~/work/setupMaezono/us
% du
```

It will respond with several lines on the '**standard output**' (usually it means a display). Then, try

```
% du > temp
```

The command, this time, does not provide any response on the standard output. Then, check the content of the file 'temp' using 'cat' or 'more'. You will find the content is the same as you saw on the standard output when executed solely 'du'. By the form, "[command] > [file name]", the responding texts of the command will be written on the file specified. This function is described as 'the output is redirected to the file'.

As you see, there is no need to handwrite or cut-and-paste the computer's output to keep track of it, but rather to redirect it and save it to a file. However, you may be wondering, 'The redirect save everything even when only the limited part is required to be recorded'. For this, we will introduce some useful **text processing** commands such as "grep/awk/sed" later (Sect. 3.4). By combining them with 'pipe', you can 'take note' by the command operation very efficiently. Once you have mastered these functions, you will have such a convenient life that it will be ridiculous to use paid Excel for processes that you used to use Excel for.

Next, execute

```
% echo 'student'
```

You will see 'student' as responded on the display. The command, "echo '[string]'", just returns the given string to the standard output.[21] Then, try

---

[21] When you first learn this command, you may wonder what it is used for, but it is a rather useful when using 'script' technique, as described later.

```
% echo 'student' > temp
```

You can confirm by 'cat' that the previous contents of temp have been **overwritten** and only "student" is left there. Then, try

```
% du >> temp
```

and check the contents of 'temp'. You will see that the 'du' output is written after the line of "student". We understand the difference that ">" overwrites while ">>" appends without overwriting.

Redirect can be used not only with ">" for output, but also with "<" for input. To see it, try the command "**bc**", which is the mnemonic of 'basic calculator'. When executed, it will be waiting for your input. Type '3+5' and press enter-key to get '8' as the answer. Typing '5/3' as the next input, you will get '1' being unsatisfactory accuracy. Then, type 'scale=3' and again try '5/3' (you can use up-arrow key for redo). Then you will get '1.666'. The 'scale=3' specifies to use three decimal places. Once getting used to using it, 'bc' is very convenient since cut-and-paste does not result in garbled characters or font takeover when you copy and paste the results to record on text editors. 'Ctrl+D' will terminate 'bc', and bring back the normal prompt.

Now, execute

```
% echo '3+5' > temp
```

to confirm that 3+5 is written to 'temp'. Next, execute

```
% bc < temp
```

to confirm that '8' is returned. By using "[command] < [file name]", you can 'feed' the contents of the file as the input of the command. Then, try

```
% echo 'scale=3' > temp
% echo '5/3'      >> temp
% bc < temp
```

to confirm to get '1.666'. See the content of 'temp' by 'cat' as well.

Next, try

```
% bc < temp > out
```

and understand what happened by checking the content of 'out' using 'cat'. If one does not know about the redirect, 'bc < temp > out' would give the impression of 'temp enclosed in parentheses' Instead, if you can read it as "the content of 'temp' is given to 'bc', and its output is redirect to 'out'", then you have been upgraded from the most beginner.

9.2.3 Activating Alias Automatically

By the way, the "source .alias" operation described in Sect. 9.2.1 is a little cumbersome in the sense that you have to execute it every time you start up a terminal. You will find a file named ".bashrc" on your home directory. Let's see the content of it:

```
% cd
% cat .bashrc
   ...
    source ~/work/setupMaezono/bash_alias
```

You can see that the file activates the alias by 'source' command at the last line. In Ubuntu, when you launch the terminal, the commands listed in the ".bashrc" are automatically executed by its convention. This last line was set up prior to this appendix using 'echo' and 'redirect' in the main text Sect. 2.4.3. If you have read through this appendix up to here, you should be able to understand what was done at the main text. This last line automatically activates a set of aliases that will be useful for this course whenever you log in. In addition to enabling aliases, you can add any commands that you want to execute automatically after login.

By the way, the 'bash_alias' has always been enabled whenever the terminal is launched, but it has been explained that aliases do not work without executing 'source .alias'. This apparent discrepancy can be understood as explained below, due to the fact that we took 'enw' and 'cup' as alias examples: While the alias file activated by 'source .alias' is

```
/Users/maezono/setupMaezono/example_alias
```

the one automatically activated by '.bashrc' is

```
/Users/maezono/setupMaezono/bash_alias
```

Since the examples, 'enw' and 'cup', are included only in 'example_alias', these cannot be activated only when 'source .alias' is executed. So far, we used 'example_alias' for educational purposes, but from now on, we will use 'bash_alias' (which is always automatically activated).

That's all for the minimum instructions for Linux required for this course. Return back to Sect. 3 in the main text to continue the tutorial.

References

1. https://en.wikipedia.org/wiki/Unix (URL confirmed on 2022.11)
2. https://en.wikipedia.org/wiki/Vi (URL confirmed on 2022.11)

Appendix B: Supplementals to the Tutorial

10

Abstract

To keep the tutorials in the main text minimum as possible, the rest of the tutorials are placed in this appendix as supplementary information. Sect. 10.1 discusses how to construct a k-point path explained in the main text Sect. 3.6.1. The second half, Sect. 10.2, describes how to improve convergence of self-consistent loops, as introduced from the main text Sects. 4.1.1, 4.1.1 and 5.3.5.

10.1 Generating k-Point Path

To depict band dispersions, a user has to provide a one-dimensional path in the first Brillouin zone of the k-space, that corresponds to the symmetry of the target crystal. The k-point path is connecting high symmetry points in the Brillouin zone. Useful tools to assist the preparation of the k-point paths are available such as using GUI-based applications (e.g. 'Xcrysden' [1]) or Python-based libraries[1] (e.g. 'pymatgen' [2]). This section explains the method using the web application "xFroggie" [3] appeared in Sect. 3.5.4.

Visiting the web site of 'xFroggie',

```
https://xfroggie.com
```

and one can see the hypertext 'Generate k-path for band structure plot'. Clicking it, and choose 'scf.in' for '1. Select file to read geometry'. Then, for '2. Select file format', select 'Quantum espresso in file' followed by clicking 'Generate K-path'.

---

[1] Do not to confuse **Python library** with "**Python scripts**". A Python library is an "assembly of useful subroutines that can be called from Python scripts".

One will get the output below, providing a path in k-space connecting high-symmetry points:

```
Line_mode KPOINTS file
50
Line_mode
Reciprocal
0.0 0.0 0.0 ! \Gamma
0.5 0.0 0.0 ! M

0.5 0.0 0.0 ! M
0.3333333333333333 0.3333333333333333 0.0 ! K

0.3333333333333333 0.3333333333333333 0.0 ! K
0.0 0.0 0.0 ! \Gamma

0.0 0.0 0.0 ! \Gamma
0.0 0.0 0.5 ! A

0.0 0.0 0.5 ! A
0.5 0.0 0.5 ! L

0.5 0.0 0.5 ! L
0.3333333333333333 0.3333333333333333 0.5 ! H

0.3333333333333333 0.3333333333333333 0.5 ! H
0.0 0.0 0.5 ! A

0.5 0.0 0.5 ! L
0.5 0.0 0.0 ! M

0.3333333333333333 0.3333333333333333 0.0 ! K
0.3333333333333333 0.3333333333333333 0.5 ! H
```

The above contents give a k-point path connecting $(0, 0, 0) \to (1/2, 1/2, 0) \to (1/3, 1/3, 0) \cdots$. The labels "$\Gamma$" etc. given in the comment lines are the conventional names of high symmetry points in the k space [4]. Following the above from the top, we can see that the path is connecting $\Gamma \to M \to K \to \Gamma \to A \to L \to H \to A$, but the paths, $L \to M$ and $K \to H$, are given separately, not connected each other. These are the ones that could not be connected by the software, so the user has to connect them into a single path by hand, as $\Gamma \to M \to K \to \Gamma \to A \to L \to H \to A \to L \to M \to K \to H$.

10.2 Improving SCF Convergence

Although not included in the main body of Chap. 4, there are often situations where convergence in SCF calculations gets to be difficult, as mentioned in Sect. 5.3.4. The target system treated in the main body is an insulator, for which case, the convergence is generally easy. For metallic systems, usually the convergence becomes more difficult, and one may be faced with a case where the convergence is never achieved by the default settings of calculations. This section explains how to cope with such cases using **smearing** technique and related concepts.

10.2.1 Sloshing and Smearing

In the case of insulators, occupied and unoccupied orbitals are clearly distinguished. [5] As a matter of course, the occupied orbitals are always occupied. For the metallic systems, however, the practical situation for simulations is not so simple: For the systems, the band continuum is occupied upto the middle where Fermi level is located. If we represent the situation by a discrete grid with a finite number of electrons, there are vast number of equivalent possibilities on which orbitals are chosen as occupied ones (Fermi **degeneracy**). We have to take one of the concrete occupying configurations which is an approximated representation of the partially occupied continuum.

In such a case, we may encounter the following phenomenon called **sloshing**: At some SCF step, an orbital is prepared as unoccupied, but a consequent SCF loop concludes that the orbital is occupied. At the next step, the orbital is assumed as occupied but the consequent SCF loop concludes it as unoccupied ... This oscillatory determination lasts forever with the results not converging very well It's just like the orbits give each other away, saying, "You go first to the occupied orbit, please", and the final seating is hard to be determined soon.[2]

Because the occupied/unoccupied is treated clearly as '0' or '1', the "occupied card" is passed around among "orbital members" and does not converge. So, instead of making it clear as '0 or 1', allowing the blurring expression like "it looks 70% occupied" etc. would improve the oscillatory situation. This corresponds to replacing the Fermi distribution function with the sharp edge of step function at zero temperature into that with smoothed edges. It is just like that "If the members are edgy and have strong personalities, there will be no compromise and nothing will be decided even if they discuss, so break down the edges of the members".

To break down the edge, intuitive ways would be to use Fermi distribution at finite temperature, or to put blurring with a Gaussian. For more efficient convergence and less bias, there are various **smearing** schemes have been developed by experts. These schemes can be specified in the input file to improve SCF convergence when required.

---

[2] When you look at a bathtub on a ship, its water surface at both ends takes the alternating rise and fall in response to the rocking of the ship, that is the original meaning of 'sloshing'.

'Smearing' is a term to describe a situation where water clings to the windshield of a car when you run the windshield wipers that has no water repellent at all. Taking the image of sloshing as "a low-viscosity liquid causing wavy surface", then the "smearing" suppresses the wavy motion by adding viscosity to the liquid.

With the above concept, we use a smearing scheme to promote and improve convergence against "difficulty in SCF convergence due to sloshing". Users can specify in the input file which smearing scheme to use and the parameters associated with the scheme (e.g., blur width for edge breaking). In the next section, we will work on an example with the scheme. The larger the smearing works, it corresponds to a stronger "viscosity to distribution variation". If it is too viscous, the distribution will stick to the point where it does not converge, and the distribution will stop updating, giving biases. We will learn a procedure how to choose proper strength of smearing so that not to get such biases.

10.2.2 Convergence Controlled by Smearing Parameter

Using an aluminum metal crystal with a FCC (face-centered cubic) lattice as an example, we learn the handling of smearing and the appearance of bias. We will compare two different smearing schemes [Gaussian smearing and Methfessel-Paxton (MP) smearing] [6] and see how the converged energy value are affected by the choice of parameters. We also see how the required iteration cycles for convergence affected by smearing width (blur width).

Prepare a working directory named "smearing" under the "~/work". From this location, let's copy the aluminum pseudopotential from the downloaded materials as follows:

```
% mkdir -p ~/work/smearing
% cd ~/work/smearing
% cp ~/work/setupMaezono/inputFiles/05smearing/Al.pbe-n-kjpaw_psl.1.0.0.UPF .
```

Next, prepare the input file "smear.in" for the SCF calculation as follows:

```
&control
    calculation='scf'
    pseudo_dir='./'
 /
&system
    ibrav=2, celldm(1) =7.50,
    nat=1, ntyp=1,
    ecutwfc =15.0,
    occupations='smearing',   ! (1)
    smearing='target1',       ! (2)
    degauss=target2           ! (3)
 /
&electrons
 /
```

```
ATOMIC_SPECIES
 Al   26.98 Al.pbe-n-kjpaw_psl.1.0.0.UPF
ATOMIC_POSITIONS (alat)
 Al 0.00 0.00 0.00
K_POINTS {automatic}
 4 4 4 1 1 1
```

The line (1) specifies the application of the smearing method with the concrete scheme specified at the line (2). The smearing parameters are specified at (3). We leave the specifications (2) and (3) as the string "target1/2", respectively, that are replaced by a script with a loop structure, "smear.sh", as described in Sect. 4.3.1:

```
% cat smear.sh
    #!/bin/bash
    for target1 in gauss mp;
    do
    for target2 in 0.05 0.02 0.01 0.005 0.002 0.001 0.0005 0.0002 0.0001;
    do
        base=$target1'_'$target2
        cat smear.in | sed -e "s#target1#$target1#g" \
            -e "s#target2#$target2#g" > $base.in
        pw.x < $base.in | tee $base.o
    done
    done
```

For "target1" in line (2) of "smear.in", the keywords, 'gauss' or 'mp', are substituted, those corresponding to the two different smearing schemes. For "target2" in line (3), it is replaced by nine different values (in Ry), being the smearing width. The loop structure of the script performs the 18 calculations one after another (it will take around two minutes) by executing

```
% bash smear.sh
```

After the calculations are completed, we first check the SCF convergence by the usual routine work using 'grep "total energy"'.

Next, let's look at the dependence of the converged energies on the choice of scheme/parameters as

```
% grep ! gauss_*.o
    gauss_0.0001.o:! total energy   =      -39.49607685 Ry
    gauss_0.0002.o:! total energy   =      -39.49608743 Ry
    ...
    gauss_0.05.o  :! total energy   =      -39.50248078 Ry

% grep ! mp_*.o
    mp_0.0001.o:!    total energy   =      -39.49607156 Ry
    ...
    mp_0.05.o  :!    total energy   =      -39.49848655 Ry
```

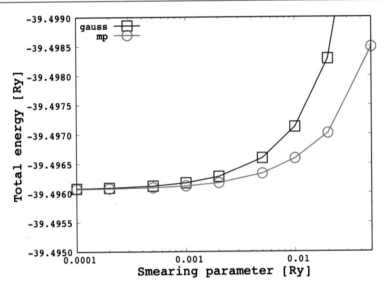

Fig. 10.1 The dependence of the total energy (vertical axis) on the smearing width (horizontal axis) is shown. The converged energy value toward the left side is taken as a reliable estimate. The comparison between the Gaussian smearing (blue square) and the MP smearing (red circle) shows the superiority of MP in the sense that their estimations are closer to the converged values even with larger smearing width, corresponding to less bias

These results are summarized as a plot shown in Fig. 10.1.

Next, let's see how the convergence is accelerated by extracting the CPU time[3] as

```
% grep WALL *.o| grep PW
      gauss_0.0001.o: PWSCF :  2.36s CPU    2.54s WALL
      ...
      gauss_0.02.o:   PWSCF :  2.08s CPU    2.39s WALL
      gauss_0.05.o:   PWSCF :  2.13s CPU    2.35s WALL
      --------------------------------
      
      mp_0.0001.o:    PWSCF :  2.01s CPU    2.25s WALL
      ...
      mp_0.05.o:      PWSCF :  1.96s CPU    2.08s WALL
```

As increasing the width from 0.0001 to 0.05 to make the smearing "more effective", the CPU time seems to be improved but quite slightly. Unfortunately, this is not a good example to watch how the smearing improves convergence because the original convergence is not so bad.

---

[3] The execution time for the calculation is written out as "WALL time" in the out file, and can be extracted as the lines containing the string, "WALL".

Rather more attention should be payed to look at the behavior shown in Fig. 10.1. Since the smearing is a prescription artificially introduced to improve convergence, the energy value approaching to the limit of the width toward zero is the reliable estimate. As explained above, if the width is increased too large, the variation of the charge distribution in the SCF gets to be too sticky, leading to a biased estimation (increased values toward the right side in this example). Comparing the two schemes, we can see that the MP scheme has less bias, even though the smearing width is taken as larger values to make the smearing effective.

When using smearing in actual operation, try several smearing widths σ and adopt the extrapolated value to $\sigma \to 0$. As we mentioned above, Fig. 10.1 is not a good example to appeal how the SCF convergence is improved or accelerated. If we take a good example, instead, then it gets to be difficult to illustrate clearly how the energies converged toward $\sigma \to 0$ because in such cases where smearing works well (hard SCF convergence), it is difficult to obtain the converged energy with smaller σ. In practice, therefore, it is a bit tricky to get the extrapolated value toward $\sigma \to 0$.

10.2.3   Convergence Acceleration by Mixing and Resume Calculation

What causes sloshing in 'poor convergence problems' is the situation where "the charge density assumed in the first stage of the SCF is significantly different from the concluded charge density by the next stage" (Fig. 5.5). If we "revolutionarily update the charge distribution at once" with this large difference between the old and the new, sloshing will occur again as the form of "counterrevolution". Based on this observation, a prescription would be possible to suppress sloshing by using a 'mixed' charge density with an appropriate ratio, such as 7:3 (70% old density mixed with 30% new one). This updates the density with more mild manner preventing from the sloshing, called **mixing of charge density update**. The ratio can be specified in the SCF input file.

In the practical projects, the convergence of the SCF calculation may sometimes take more than a week in severe cases. In such a case, it is risky to leave the calculation running for a week or longer. For example, if the sudden blackout occurs on the fifth day, the entire five days of calculations will be wasted. Also, when using a shared computer resources such as a supercomputer, there is usually a **job time limit** restricting a job running maximum within 24 h or 2 days. If you understand the mechanism of the SCF convergence, namely, "gradually equilibrating the charge density distribution", you can understand that even in such a case, you can save the charge distribution that has reached the halfway point toward equilibrium, terminate the calculation once, and **resume** the calculation by loading the previous distribution as the **initial guess** for the next calculation.

In the input file shown in Sect. 3.3.1, there is a setting named "restart_mode= 'from_scratch'". If you change this setting to 'restart', the calculation mode is set as resuming, namely, reading the initial distribution from the specified directory and restarting the calculation. Setting it as 'from_scratch' means, conversely, that the calculation starts with the initial distribution newly generated from scratch.

It is often the case that beginners do not know the strategy using the "resume calculation": Trying 'from_scratch' not converging after one day, then trying again 'from_scratch' for three days, and then trying again 'from_scratch' for a week, ... This is a huge waste of time and computing resources. It is important to understand how the SCF calculation works, and to imagine that your charge density is gradually equilibrating towards convergence on the computer. Upon such understanding, the above mistakes never happens, and you can be regarded to be grown up from a beginner to a practitioner.

References

1. http://www.xcrysden.org (URL confirmed on 2022.11)
2. https://pymatgen.org (URL confirmed on 2022.11)
3. https://xfroggie.com (URL confirmed on 2022.11)
4. "Kittel's Introduction to Solid State Physics" C. Kittel, John Wiley & Sons Inc (2018/7/9) ISBN-13:978-1119454168
5. "Solid State Physics", Neil Ashcroft, N. Mermin, Brooks/Cole (2021/4/8) ISBN-13:978-0357670811
6. https://www.quantum-espresso.org (URL confirmed on 2022.11)

Appendix C: Band Theory

11

Abstract

Appendix Sects. 11.1–11.4 is derived from the main text Sect. 3.4. The construction of the contents is a bit different from the conventional textbook explaining the band theory. The motivation of this reconstructed explanation is to answers the naive question made from practitioners asking why we want to calculate the band dispersion. We begin with the overall logic flow explained in Sect. 11.1. Individual concepts such as mode separation, 'symmetry and commutation relation' etc. used in the logic flow are explained in the subsequent sections. Beginners who are not familiar with these concepts can read through these sections, and getting back to Sect. 11.1, and trying again to read through. The latter part starting from Sect. 11.5 treats the slightly more advanced topic about the phase of the wave function. Appendix Sect. 11.5 is derived from the main text Sect. 6.1.2, while Appendix Sect. 11.5.3 from the main text Sect. 4.2.3.

11.1 Overview

11.1.1 Fundamental Information to Understand Materials Response

'Utilization of materials properties' mostly refers to the use of 'the response properties' of materials. In the picture of 'stimulus and response', a material system receives energy from outside and responds by changing its internal state. Suce responses include "electrons receiving energy and excited to uppper levels", "spins changing their directions", "impurities starting to move", or "lattice vibrations induced" etc. In these responses, the degree of freedom that can change the state is where the entropy exists. The first step in understanding the properties of materials is to know what degrees of freedom exist in the material to be considered, and the corresponding experiment would be the specific heat measurement.[1]

The possible forms of material response is defined by the available transition patterns among the states of degrees of freedom which responds to the given energy as the stimulus. The fundamental information to predict the material response is, therefore, to know how dense the possible states are distributing that can receive the given energy or that can accommodate the excited electrons. Once knowing this, we can evaluate the response based on the golden rule as explained in the main text Sect. 6.2.1.

11.1.2 Mode Separation Used for the Analysis

If such transition patterns can be understood as the two-level model between the ground- to excited-state, it is not too difficult. As in the example of lattice vibrations, however, it is not so simple picture because when we agitate the state of a degree of freedom, it affects to other degrees of freedom that are coupled together. It cannot directly be connected to the picture of the two-level model.

Here we would like to remember the method of mode separation (Sect. 11.2). Complicatedly coupled vibrations can be decomposed into independent **normal modes** by using the diagonalization method. To formulate the coupled vibration, we start to write down the equation of motion by focusing on the displacement of each atom (ψ_a and ψ_b in Fig. 11.1) as naive manner, leading to a set of coupled equations. However, if we take a good linear combination, e.g., $\psi^{(\pm)} = \psi_b \pm \psi_b$, then $\psi^{(+)}$ and $\psi^{(-)}$ do not hybridize with each other to form decoupled modes (normal modes).

The prescription to find this kind of proper linear combination is the diagonalization of the matrix. The decoupled 'normal modes' found in this way can be treated as separate and independent degrees of freedom that do not mix with others. These independent modes are considered to be "solely excited from the ground- to the excited-state". The possible patterns of the excited transition of the system is then captured as the sum of these modes.

11.1.3 Symmetric Transformation of Wavefunctions

The orthodox target of electronic structure calculations is a periodic array of complexes. In such systems, there are **symmetric operations** where "the view looks the same even if the viewpoint is shifted by one period of the array", "the same view even if the viewpoint is rotated by $\pi/3$ around the c-axis. Symmetrical operations can be expressed by exchanging the coordinate labels, e.g., $x \rightarrow y$, $y \rightarrow (-x)$ (in the case of a $\pi/2$ rotation around the z-axis, for example). Such "viewpoint shifts"

---

[1] For example, from the temperature dependence of specific heat, we can find out what degree of freedom gets to be activated/deactivated.

are generally represented by coordinate transformation operators such as $\mathbf{r}' = \hat{R} \cdot \mathbf{r}$ (rotation operator) and $\mathbf{r}' = \hat{T} \cdot \mathbf{r}$ (translation operator).

When a symmetric operation such as rotation or translation is applied to a describing coordinate system, the field (wavefunction, spin, etc.) on the coordinate system also undergoes a transformation. Suppose that an symmetric operation $\hat{L}$ is represented by a coordinate transformation $\mathbf{r} \to \hat{R} \cdot \mathbf{r}$. For this operation, the transformation of the wave function is represented as

$$\hat{L}\psi\,(\mathbf{r}) = \psi\left(\hat{R} \cdot \mathbf{r}\right) \tag{11.1}$$

We will formulate the problem in the form of what transformation $\hat{L}\psi\,(\mathbf{r})$ will undergo. When a describing coordinate is transformed, each component of the field that was classified in terms of the previous coordinate gets mixed up with each other in a complicated manner, leading to poor outlook. This mixed-up situation can be decoupled by the mode separation technique as block-diagonalized to get a subgroup with three components (labeled, say, by Γ_a) or a subgroup with five components (Γ_b) etc., where the transformation $\hat{L}\psi\,(\mathbf{r})$ exchanges only the components each other within a subgroup (Sect. 11.2.3). This is the mode separation into the 'factions' that do not mix beyond the the label $\Gamma_{\alpha=a,b,...}$".[2] Henceforth, we use the index Γ applied to an electronic state $\psi\,(\mathbf{r})$, which distinguishes the mode-separated 'factions'.

11.1.4 Indexing Energy Levels of Periodic Systems

To get the distribution of the energy levels, we want to know how the eigenvalues of the Hamiltonian are indexed. If the Hamiltonian remains invariant to the symmetric operation $\hat{L}$,[3] then it is concluded that its eigenvalues ε are classified by the transformability of $\hat{L}$, indexed by Γ as introduced above (Sect. 11.3). For example, an isolated atom should not be affected by the rotation of the coordinates around the origin (that's the human perspective). As we know, the energy levels are indexed as s, p, and d etc., which are actually the indices of the mode separation Γ for the rotation operation. Since the nature should not be affected the labelling of human perspective such as p_x, p_y, etc., the transformation (namely, the relabeling of $p_x \to p_y$ etc. like changing seats) should keep the energy level, leading to the fact that the level degeneracy should exist among $p_{x,y,z}$ that transform each other.

Following this idea, let's consider what is Γ for a translational symmetry. The Γ is the classifying index with respect to the translational operation $\hat{T}_\mathbf{R}$ applied to $\psi\,(\mathbf{r})$.

---

[2] As a concrete image, you can remember the 'factions of p-orbitals' and 'factions of d-orbitals'. When some rotationally symmetric operation is applied, a p-orbital is transformed to another p-orbital (p-mode), while a d-orbital to another d-orbital (d-mode). p and d do not mix with each other (mode-separated).

[3] The fact that a system is symmetric with respect to the operation is mathematically represented as the invariance of the Hamiltonian of the system with respect to the operations.

It is sorted in terms of how the 'view' looks as the viewpoint is shifted one by one along the period of the array. That is "AAAA..." (the phase is invariant with respect to the shift, indexed as "$k = 0$"), or "ABAB..." (the phase alternating at every shift, indexed as "$k = \pi$"), or "ABCABC..." (the phase getting back at every three shifts as indexed "$k = 2\pi/3$") (Sect. 11.4.1). In the usual textbook wording, this is "how much phase angle is earned by $\psi(\mathbf{r})$ when it is advanced by one unit cell".

One might have the question, "Sure, such a classification is possible, but is it really a mode separation that does not mix with each other?". That is the role of the Bloch's theorem (Sect. 11.4.3) to ensure the affirmative answer. For the translational operation $\hat{T}_{\mathbf{R}}$, the Bloch function $\psi_{\mathbf{k}}$ used in electronic state calculations is shown to satisfy

$$T_{\mathbf{R}} \cdot \psi_{\mathbf{k}}(\mathbf{r}) = \exp[i \cdot \mathbf{k} \cdot \mathbf{R}] \cdot \psi_{\mathbf{k}}(\mathbf{r}) \tag{11.2}$$

In this formula, we can see that the diagonalizing eigenvalue (i.e., the index of the mode separation) is surely $\mathbf{k}$ which is the phase angle obtained when $\psi(\mathbf{r})$ is advanced by one unit cell $\mathbf{R}$. This is how to "taste" Bloch's theorem.

That is the overviewing explanation that the energy level of a periodic system is labelled by $\mathbf{k}$ as the mode-separated index. By the way, readers might have already learned Fourier transform somewhere. The Fourier transform is actually understood as the mode separation to decouple the inter-site coupled motions, for which the index *Gamma* corresponds to the wavenumber $\mathbf{k}$ (Sect. 11.2.4).

For the periodic systems (which reciprocal lattice is specified as $\mathbf{G}$), it is also derived that the $\mathbf{k}$ and $(\mathbf{k} + \mathbf{G})$ are equivalent in the sense of phase angle (Sects. 11.4.2, 11.4.3). The meaning of the 'equivalence' is in the same sense as 'θ and $\theta + 2\pi N$ (N; integer) are equivalent'. For the case of θ, we can therefore use the representative value specified within the unit circle. This unit circle corresponds to the first Brillouin Zone in the periodic system (Sect. 11.4.2). In Sect. 11.4, it is explained 'How and Why a reciprocal lattice is introduce?' as well as 'Why the reciprocal lattice is defined in that way?'.

In summary,

- Distribution of energy levels provides the essential information to understand the response of the material.
- Indexing for the energy levels is known to be able to be classified by how the states corresponding to the levels transform with respect to the symmetric operations.
- For a periodic system, the indexing is made by the wave vector $\mathbf{k}$, which is the index of the mode separation with respect to the translational symmetry. This $\mathbf{k}$ can be represented within the first Brillouin zone.
- For the index, a point $\mathbf{k} = \bar{\mathbf{k}}$ within the first Brillouin zone, each index $\bar{\mathbf{k}}$ is accompanied with a further index b as $\varepsilon(b, \bar{\mathbf{k}})$ ($b = 1, 2, \ldots$), that is originated from the mode separation with respect to the spatial symmetry for the complex at each lattice point.
- Each discrete level labelled by b can form continuous dependence on $\mathbf{k}$, which can be plotted along a path in the $\mathbf{k}$-space. That is the band dispersion diagram, taking the role of the visualized 'list of the energy levels possible for the system'.

'Visualized profiling' of a periodic system is a 'graphical representation of the energy level distribution, $\varepsilon\,(b, \mathbf{k})$'. That is the meaning and significance of the band dispersion diagram as shown in Fig. 3.10 (Sect. 3.6), answering the question, 'Why we want to evaluate it'.

11.1.5 Concept of Elementary Excitations

In discussions on materials science, various concepts of elementary excitations appear, such as polarons, polaritons, excitons, plasmons, magnon, rotons [1], holons, spinons [2] etc. For example, a polaron is explained as 'a quasiparticle based on an electron interacting with a lattice vibrations'. However, for a beginner, such explanations are not satisfying. Though each sentence itself is understandable, but it never answering the questions like, 'why do we focus on polarons in this system while it is excitons for another system? Both systems are having the same set of degrees of freedom...'.

Although every system has various possible degrees of freedom, there would be an established consensus for each problem to be analyzed that what is the primary degree of freedom to dominate the phenomena. The first step of the formulation is to perform the mode separation of the coupled variables applied only to the primary degrees of freedom. The mode-separated vibrations are then typically processed by the canonical quantization, getting reduced to the quantum statistical mechanics of N harmonic oscillators with ascending/descending operators [1]. During the mode-separation procedure, one gets the list of effective masses and frequencies for each separated mode. When this procedure is applied to lattice vibrations, electromagnetic waves, vibrations of spin directions, etc., it gets to phonons, photons, magnons etc.. The flow of the procedures applied to the phonons is a prototype [1] for other elementary excitations.

> **Important**

This is the reason why we are forced to study quantum statistical mechanics of harmonic oscillators [1]. Often, the lack of this explanation discourages students, as they feel that '... it's just an exercise with ideal models, which have nothing to do with real material systems...'. It is common strategy to pick upto second order by expanding any degree of freedom with respect to the tiny deviation (harmonic approximations). It is also common strategy to make mode-separation applied to the coupled degree of freedom with harmonic approximation, getting a set of independent harmonic oscillators. As such, any problems reduce to the Quantum/Statistical mechanics of the harmonic oscillators, so this is a general framework that everyone should study.

In some cases, coupled oscillations of different degrees of freedom are considered to be the primary degrees of freedom. They are, for example, the coupling between electromagnetic fields and lattice vibrations, or the coupling between the electrons and lattice vibrations. They are then decoupled by the mode separation into mutually independent elementary excitations. Depending on the combination between what is considered as primary degrees of freedom, it is called polarons, polaritons, excitons etc. Again note that which degrees of freedom are taken as primary is based on the consensus formed for each target phenomenon.

Mode-separated elementary excitations are a picture that is valid within the harmonic approximation, but if the power expansion is justified, higher-order effects above the third order can be treated in the perturbation manner. The effect can be described as a scattering between the elementary excitations (basic picture within the harmonic approximation), which makes the lifetime for the excitations finite [3].[4]

11.1.6 Band Dispersion Gives What Information?

In response to the simple question, "Why do we calculate and evaluate band dispersion?", I have explained in the main text that the dispersion is just like the output of "fish population detector" to capture where is the possible energy levels to evaluate the golden rule Eq. (6.1). In this section, I will explain more concretely what kind of information the practitioners can extract from the calculated dispersions.

By looking at the band dispersion, the first thing to be checked is the presence or absence of a gap and the gap value, which tells us at what energy the electrons under the Fermi level can be excited. This determines whether the system is a metal or an insulator [4].

> **! Attention**
>
> When a system can be excited at infinitesimal values due to the zero energy gap, it is a metallic system. For the finite energy gap, there is the problem of the bandgap underestimation as mentioned at the end of Sect. 3.7 (a more detailed explanation is given in Chap. 12). For a more comprehensive view to distinguish metals and insulators, an explanation is given in Sect. 11.5.2.

When the system has a finite bandgap, then we focus the magnitude of the gap compared to the energy range applied to the system to be excited. For example, whether the photo-excitation energy, $\Delta\varepsilon = \hbar\omega$, can overcome the bandgap or not corresponds to whether the system absorbs light or not, namely, transparent or not. In this case, depending on whether the highest valence band position and the lowest conduction band position are at the same $\mathbf{k}$ ('direct transition') or not ('indirect transition'), the behavior when the system is excited by light differs. As such, whether

---

[4] Within the harmonic approximation, the lifetime is infinite, corresponding to the fact that the mode-separated solutions (diagonalized vectors) are the eigenstate of the harmonically approximated Hamiltonian.

direct or indirect transition is another point to be checked first when the band dispersion is evaluated.

In the diagram, there are some band branches with larger dispersion, while some are with little dispersion.[5] It is important for one to be trained up so that you can imagine more itinerant electrons when you see a branch with larger dispersion. This is usually explained by using the tight-binding model [4]: If we solve the model with the hopping amplitude to neighboring sites as t, we can derive that the bandwidth is proportional to t, which leads to the understanding that the larger hopping amplitude gives the larger dispersion. The larger dispersion with the larger curvature therefore corresponds to the more itinerant electrons. For this reason, electrons and holes in valleys with larger curvature are called "lighter carriers" (easy to move) and those in valleys with smaller curvature called "heavy carriers" (won't move) [1].

In any under graduate course on semiconductors, one would learn about the impurity level [5]. The impurity level is drawn a little below the conduction band with such a qualitative explanation that 'electrons trapped in impurities can be pulled out by a tiny amount of energy to hop up into the conduction band'. When you perform an *ab initio* DFT calculation for a crystal structure with impurities introduced, you will obtain the DOS showing the impurity level actually appearing beneath the Fermi level [6]. Comparing the band dispersions for the system with and without impurities, you will find a new branch with almost no dispersion appeared corresponding to the introduction of the impurities. For such a branch with little dispersion, all the states on the branch are projected within a narrow energy range to form a sharp peak in DOS.

As mentioned above, a branch with little dispersion is interpreted as a localized state, which is heavy to move. The sharp DOS peak appeared by the impurity doping is, therefore, representing that the electrons are trapped around an impurity to be localized. This would be a full explanation of the impurity level as an update of the qualitative explanation given in undergraduate course. Suppose that the impurity peak in DOS loses its sharpness when the system is subjected to additional conditions, such as elemental substitution or pressure application. It corresponds to that the branch of the impurity level gets dispersed, as interpreted that the trapped electrons at the impurity gets released to start hopping by the application of the additional conditions. As another consequence of the additional conditions, if the impurity level goes into the valence or conduction band, it is interpreted that the localized electrons start to be mixed up with other medium-electrons to get indistinguishable. For doped systems, therefore, it is the point to be focused on DOS diagram whether peaky states appears or not [7].

When DOS is calculated for a one- or two-dimensional free electron model, characteristic shapes and singularities appear in each dimension, known as van Hove singularities. This fact can be used as a reference to analyse the DOS for practical

---

[5] When a band has larger dependence on k, the branch bends largely, that is called 'larger dispersion'.

systems. If the practical DOS includes the characteristic shape of the van Hove singularities [3], it would imply that the system has some low-dimensional nature in its conduction.

11.2   Mode Separation and Diagonalization

11.2.1 Introduction of Mode Separation

Considering a spring-mass system (K and m denoting the spring constant and mass),

$$m\frac{d^2}{dt^2}\cdot\psi\left(t\right)=-K\cdot\psi\left(t\right)\;, \tag{11.3}$$

the natural frequency, $\omega=\sqrt{K/m}$, is the quantity to characterize the system, working as a mode index.

Let us consider a coupled vibrating system as shown in Fig. 11.1. Equations of motion are

$$m\frac{d^2}{dt^2}\cdot\psi_a=-K\cdot\psi_a-k\cdot(\psi_b-\psi_a)$$
$$m\frac{d^2}{dt^2}\cdot\psi_b=-K\cdot\psi_b-k\cdot(\psi_a-\psi_b)\;. \tag{11.4}$$

Two modes ψ_a and ψ_b (deviations from each equilibrium position) are coupled each other and apparently behave in complex manner. Though it is heuristically, taking the sum and difference of ψ_a and ψ_b, we get

$$m\frac{d^2}{dt^2}\cdot(\psi_a+\psi_b)=-K\cdot(\psi_a+\psi_b)$$
$$m\frac{d^2}{dt^2}\cdot(\psi_a-\psi_b)=-K\cdot(\psi_a-\psi_b)-2k\cdot(\psi_a-\psi_b)\;. \tag{11.5}$$

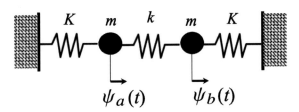

Fig. 11.1 A simple model of coupled vibrations where each mass m bound to the wall with the spring constant K is coupled with each other by a spring k. $\psi_{a,b}\left(t\right)$ denotes the deviation from each equilibrium position

Fig. 11.2 Such a level splitting diagram, which is often seen [8], is a graphical representation of the mode separation where each original mode contributing to hybridization gets split into different eigenvalues

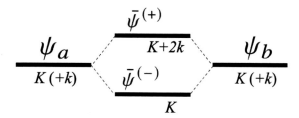

Then, defining[6]

$$\bar{\psi}^{(\pm)} := \psi_a \pm \psi_b , \tag{11.6}$$

the equations of motion get to

$$m\frac{d^2}{dt^2} \cdot \bar{\psi}^{(+)} = -K \cdot \bar{\psi}^{(+)}$$

$$m\frac{d^2}{dt^2} \cdot \bar{\psi}^{(-)} = -(K + 2k) \cdot \bar{\psi}^{(-)} . \tag{11.7}$$

Here the two equations with respect to $\bar{\psi}^{(\pm)}$ are independent, not coupled each other anymore. The decoupled variables, $\bar{\psi}^{(\pm)}$, are the separated modes. By this **mode separation**, the original modes $\psi_{a,b}$ with the characteristic constants, K and k, are separated into $\bar{\psi}^{(+)}$ with K and $\bar{\psi}^{(-)}$ with $(K + 2k)$, which is schematically depicted as in Fig. 11.2.

In this example, a proper linear combination, $\bar{\psi}^{(\pm)}$, to achieve the mode separation is given intuitively, However, we wondered if we could construct such proper mode separation not heuristically, but by a general prescription. The prescription is actually corresponding to the matrix diagonalization, as explained in the next section.

11.2.2 Prescription of Mode Separation

Denoting $\lambda = -\omega^2$ for the $\omega = \sqrt{K/m}$ of the equation of motion of the spring-mass system, Eq. (11.3), it can be written in the form of a linear eigenvalue problem as

$$\hat{A} \cdot \psi = \lambda \cdot \psi , \tag{11.8}$$

where we denote the linear operator, $\hat{A} = d^2/dt^2$. For the dynamics driven by $\hat{A}$, let us denote the mode mixing corresponding to Eq. (11.4) as

$$\hat{A} \cdot \psi_l = a_1^{(l)} \cdot \psi_1 + a_2^{(l)} \cdot \psi_2 + \cdots + a_M^{(l)} \cdot \psi_M = \sum_{m=1}^{M} a_m^{(l)} \psi_m , \tag{11.9}$$

---

[6] The bar put on the variables implies 'the mode separated' hereafter.

where $\{\psi_m\}_{m=1}^M$ are the modes hybridizing each other (before the 'mode-separation'). As $\psi^{(\pm)}$ appeared in the previous subsection, let us suppose a linear combination

$$\bar{\psi} = \sum_{l=1}^{M} c_l \cdot \psi_l , \qquad (11.10)$$

which realizes the mode-separation, namely it satisfies

$$\hat{A} \cdot \bar{\psi} = \lambda \cdot \bar{\psi} , \qquad (11.11)$$

without any hybridization with other modes. Substituting the assumption, Eq. (11.10), into Eq. (11.11), and employing the notation, $a_m^{(l)} = a_{ml}$, we get

$$\lambda \cdot \bar{\psi} = \hat{A} \cdot \bar{\psi} = \sum_{l=1}^{M} c_l \cdot \left(\hat{A} \cdot \psi_l \right) = \sum_{l=1}^{M} c_l \cdot \left(\sum_{m=1}^{M} a_{ml} \psi_m \right)$$

$$= \sum_{m} \left(\sum_{l} a_{ml} \cdot c_l \right) \psi_m . \qquad (11.12)$$

About the expanding coefficients with respect to ψ_m, we can equate a function and a vector as

$$\bar{\psi} = \sum_{m=1}^{M} c_m \cdot \psi_m \quad \longleftrightarrow \quad \begin{pmatrix} c_1 \\ c_2 \\ \vdots \\ c_M \end{pmatrix} . \qquad (11.13)$$

Comparing the coefficients of any specific ψ_m in Eq. (11.12), the relation can be written,

$$\lambda \cdot \begin{pmatrix} c_1 \\ c_2 \\ \vdots \\ c_M \end{pmatrix} = \begin{pmatrix} a_{11} & a_{12} & \cdots & a_{1M} \\ a_{21} & \cdots & & \\ \vdots & & & \\ a_{M1} & \cdots & & a_{MM} \end{pmatrix} \cdot \begin{pmatrix} c_1 \\ c_2 \\ \vdots \\ c_M \end{pmatrix} , \qquad (11.14)$$

as a matrix-vector product form.

We notice that this corresponds to solving a matrix eigenvalue problem (diagonalization of a matrix) for the coefficient matrix for 'the equation before the mode-separation' (Eq. (11.9)). Taking the example of Eq. (11.4) rewritten as

$$\frac{d}{dt} \begin{pmatrix} \psi_a \\ \psi_b \end{pmatrix} = \begin{pmatrix} -K+k & -k \\ -k & -K+k \end{pmatrix} \cdot \begin{pmatrix} \psi_a \\ \psi_b \end{pmatrix} ,$$

the coefficient matrix is

$$A = \begin{pmatrix} -K+k & -k \\ -k & -K+k \end{pmatrix} = \begin{pmatrix} a_{11} & a_{12} \\ a_{21} & a_{22} \end{pmatrix}.$$

To diagonalize the matrix to solve the eigenvalue problem, we start with getting eigenvalues λ as

$$\begin{vmatrix} (-K+k)-\lambda & -k \\ -k & (-K+k)-\lambda \end{vmatrix} = 0$$

$$\therefore \quad [(-K+k)-\lambda]^2 = k^2$$

$$\therefore \quad (-K+k)-\lambda = \pm k$$

$$\therefore \quad \lambda = -K+k \pm k$$

$$\therefore \quad \lambda^- = -K+2k \quad, \quad \lambda^+ = -K.$$

getting two eigenvalues for Eq. (11.14). The eigenvectors in Eq. (11.14) correspond to the expanding coefficients for the mode-separated expression Eq. (11.10), being

$$\bar{\psi}^\pm = c_1^\pm \cdot \psi_a + c_2^\pm \cdot \psi_b$$

in this case. For each case of $\lambda^\pm$, Eq. (11.14) becomes

$$\begin{pmatrix} -K+k-(-K+2k) & -k \\ -k & -K+k-(-K+2k) \end{pmatrix} \cdot \begin{pmatrix} c_1^- \\ c_2^- \end{pmatrix} = 0$$

$$\begin{pmatrix} -K+k-(-K) & -k \\ -k & -K+k-(-K) \end{pmatrix} \cdot \begin{pmatrix} c_1^+ \\ c_2^+ \end{pmatrix} = 0$$

$$\therefore \quad \begin{pmatrix} -k & -k \\ -k & -k \end{pmatrix} \cdot \begin{pmatrix} c_1^- \\ c_2^- \end{pmatrix} = 0 \quad, \quad \begin{pmatrix} k & -k \\ -k & -k \end{pmatrix} \cdot \begin{pmatrix} c_1^+ \\ c_2^+ \end{pmatrix} = 0,$$

leading $c_1^\pm = \pm c_2^\pm$. This leads to

$$\bar{\psi}^\pm = \psi_a \pm \psi_b,$$

as the separated modes, giving the same conclusion as shown in Eq. (11.6) [heuristically obtained before].

In short, the prescription to get the coefficients of a linear combination, Eq. (11.10), to realize the mode separation is achieved by diagonalizing the coefficient matrix for the equation before the mode-separation, (Eq. (11.9)), as written in Eq. (11.14) as a matrix eigenvalue problem. The resultant components of the eigenvectors, $\{c_l\}_{l=1}^M$, give the expanding coefficients to form the separated modes.

Since total number of the modes to be hybridized, M, in Eq. (11.9) is the dimension of the vector, we obtain M eigenvalues, $\{\lambda_\Gamma\}_{\Gamma=1}^{M}$, and M eigenvectors corresponding to each of them,

$$\mathbf{u}^{(\Gamma=1,\dots,M)} = \begin{pmatrix} c_1^{(\Gamma)} \\ c_2^{(\Gamma)} \\ \vdots \\ c_M^{(\Gamma)} \end{pmatrix} \quad\longleftrightarrow\quad \bar{\psi}^{(\Gamma)} = \sum_{m=1}^{M} c_m^{(\Gamma)} \cdot \psi_m . \tag{11.15}$$

These M-fold,

$$\bar{\psi}^{(\Gamma)} = \sum_{m=1}^{M} c_m^{(\Gamma)} \cdot \psi_m \quad (\Gamma = 1, \dots, M) , \tag{11.16}$$

are the independent modes (mode-separated ones) that are not hybridized anymore, called **normal modes**.

The $\psi^{(\pm)}$ appeared in the previous subsection are actually the normal mode for Eq. (11.4), which is obtained as the eigen vectors of the prescription in the form of matrix eigenvalue problem (i.e. matrix diagonalization).

11.2.3 Mode Separation for Symmetric Operations

Denoting the transform of the describing coordinates driven by a symmetric operation as $\mathbf{r}' = \hat{R} \cdot \mathbf{r}$, let us consider the corresponding transform of a field $\psi(\mathbf{r}) \to \hat{L}\,\psi(\mathbf{r})$,

$$\hat{L}\,\psi(\mathbf{r}) = \psi\left(\hat{R} \cdot \mathbf{r}\right) . \tag{11.17}$$

The field can be expanded as

$$\psi(\mathbf{r}) = c_1 \cdot \chi_1(\mathbf{r}) + c_2 \cdot \chi_2(\mathbf{r}) + \cdots + c_M \cdot \chi_M(\mathbf{r}) , \tag{11.18}$$

by the basis functions, $\{\chi_j(\mathbf{r})\}$, with well-known behavior about the transform.[7] Then,

$$\hat{L}\,\psi(\mathbf{r}) = c_1 \cdot \chi_1\left(\hat{R} \cdot \mathbf{r}\right) + c_2 \cdot \chi_2\left(\hat{R} \cdot \mathbf{r}\right) + \cdots + c_M \cdot \chi_M\left(\hat{R} \cdot \mathbf{r}\right) . \tag{11.19}$$

Let the transform of each basis function be described as

$$\hat{L}\chi_j(\mathbf{r}) = \chi_j\left(\hat{R} \cdot \mathbf{r}\right)$$
$$= a_1^{(j)} \cdot \chi_1(\mathbf{r}) + a_2^{(j)} \cdot \chi_2(\mathbf{r}) + \cdots + a_M^{(j)} \cdot \chi_M(\mathbf{r}) . \tag{11.20}$$

---

[7] E.g., Fourier expansions for periodic system, spherical harmonic expansions for spherically symmetric systems, Bessel expansions for uniaxial symmetry etc. [9].

Although we can easily write down the above relation by the well-known knowledge on $\{\chi_j\}$, they are hybridizing each other for the operation $\hat{L}$, leading a poor visibility situation (before mode-separation).

Here, we assume a linear combination,

$$\bar{\psi}(\mathbf{r}) = \bar{c}_1 \cdot \chi_1(\mathbf{r}) + \cdots + \bar{c}_M \cdot \chi_M(\mathbf{r}) , \qquad (11.21)$$

satisfying the eigenfunction about $\hat{L}$,

$$\hat{L}\,\bar{\psi}(\mathbf{r}) = \bar{\psi}\left(\hat{R}\cdot\mathbf{r}\right) = \lambda \cdot \bar{\psi}(\mathbf{r}), \qquad (11.22)$$

with properly chosen coefficients $\{\bar{c}_j\}$, realizing no inter-mode hybridization (mode-separation). This linear combination is called **irreducible representation**.

To determine the proper coefficients, $\{\bar{c}_j\}$, let us substitute Eq. (11.21) into Eq. (11.22) to get,

$$\hat{L}\,\bar{\psi}(\mathbf{r}) = \sum_l \bar{c}_l \cdot \chi_l\left(\hat{R}\cdot\mathbf{r}\right) = \sum_l \sum_m \bar{c}_l \cdot a_m^{(l)} \chi_m(\mathbf{r})$$
$$= \lambda \cdot \bar{\psi}(\mathbf{r}) \qquad (11.23)$$

Denoting $a_m^{(l)} = a_{ml}$, it leads to

$$\sum_m \left(\sum_l \bar{c}_l \cdot a_{ml}\right) \chi_m(\mathbf{r}) = \lambda \cdot \sum_m \bar{c}_m \cdot \chi_m(\mathbf{r}) . \qquad (11.24)$$

Comparing the coefficient with respect to $\chi_m(\mathbf{r})$, one gets

$$\lambda \cdot \begin{pmatrix} \bar{c}_1 \\ \bar{c}_2 \\ \vdots \\ \bar{c}_M \end{pmatrix} = \begin{pmatrix} a_{11} & a_{12} & \cdots & a_{1M} \\ a_{21} & \cdots & & \\ \vdots & & & \\ a_{M1} & \cdots & & a_{MM} \end{pmatrix} \cdot \begin{pmatrix} \bar{c}_1 \\ \bar{c}_2 \\ \vdots \\ \bar{c}_M \end{pmatrix}, \qquad (11.25)$$

as a form of matrix eigenvalue problem. This is the problem to diagonalize the coefficient matrix for the equation 'before mode-separation', Eq. (11.20).

Solving the problem, one obtains M-fold eigenvectors, $\left\{\bar{c}_j^{(\Gamma=1,\dots,M)}\right\}$, and then the mode-separated (irreducible) representations, $\left\{\bar{\psi}^{(\Gamma=1,\dots,M)}(\mathbf{r})\right\}$.[8]

---

[8] To make explanation conceptual, we didn't care about the degeneracy in eigenvalues. It is actually accompanied by the degeneracy, leading the procedure as block diagonalization, where the index Γ is classified as Γ_1 (γ_1) (e.g., s-orbital), Γ_2 ($\gamma_1,\dots,\gamma_3$) (e.g., p-orbital), Γ_3 ($\gamma_1,\dots,\gamma_5$) (e.g., d-orbital) etc.. Degenerating modes indexed as $gamma_j$ under a Γ are transformed each other within the Γ.

11.2.4 Fourier Expansion as a Mode-Separation

Let us consider a periodic system with inter-site coupling U_{ij} (i, j are site index),

$$\Phi = \sum_{i,j} \phi_i \cdot U_{ij} \cdot \phi_j \ . \tag{11.26}$$

Substituting a Fourier transform [10],

$$\phi_i = \sum_q F_q \cdot e^{iqR_i} \ , \tag{11.27}$$

we get a form,

$$\Phi = \sum_{i,j} \sum_{q,q',p} F_q \cdot U_p \cdot F_{q'} \cdot e^{iqR_i} e^{ip(R_i-R_j)} e^{iq'R_j} \ , \tag{11.28}$$

where U_p is Fourier transform of U_{ij},[9]

$$U_{ij} = \frac{1}{N} \sum_p U_p \cdot e^{ip(R_i-R_j)} \ . \tag{11.29}$$

Using the orthogonalization relation,

$$\sum_i e^{iqR_i} \cdot e^{iq'R_i} \sim \delta_{q,-q'} \ ,$$

we get

$$\Phi = \sum_{q,q',p} F_q \cdot U_p \cdot F_{q'} \cdot \delta_{q+p} \delta_{q'-p} = \sum_q F_q \cdot U_{-q} \cdot F_{-q} \ , \tag{11.30}$$

as a diagonalized form with respect to the wave number index, q.

Noticing that Eq. (11.27) is a linear combination as in Eq. (11.10), the Fourier transform is actually the mode-separation applied to the inter-site coupling.

---

[9] If the system is periodic with respect to R_i, the quantity depending on R_{ij} will be periodic correspondingly.

11.3 Index for Symmetric Operations and Energy

11.3.1 Representation of Symmetry Using Commutation Relation

Let us consider

$$\hat{H}|\phi_n\rangle = \varepsilon_n|\phi_n\rangle\,, \tag{11.31}$$

with the eigenstate $|\phi_n\rangle$ for the Hamiltonian operator $\hat{H}$. Applying a symmetric operation, $\hat{L}\,|\phi_n\rangle$, from left hand side,

$$\hat{L}\cdot\hat{H}\cdot|\phi_n\rangle = \hat{L}\cdot\varepsilon_n|\phi_n\rangle = \varepsilon_n\left(\hat{L}|\phi_n\rangle\right)\,. \tag{11.32}$$

Rewriting the left hand side as

$$(\text{LHS}) = \hat{L}\cdot\hat{H}\cdot|\phi_n\rangle = \hat{L}\cdot\hat{H}\cdot\hat{L}^{-1}\hat{L}|\phi_n\rangle = \hat{L}\hat{H}\hat{L}^{-1}\left(\hat{L}|\phi_n\rangle\right)\,, \tag{11.33}$$

Equation (11.32) leads to

$$\hat{L}\hat{H}\hat{L}^{-1}\left(\hat{L}|\phi_n\rangle\right) = \varepsilon_n\left(\hat{L}|\phi_n\rangle\right)\,. \tag{11.34}$$

If the relation

$$\hat{L}\hat{H}\hat{L}^{-1} = \hat{H} \tag{11.35}$$

holds, then

$$\hat{H}\left(\hat{L}|\phi_n\rangle\right) = \varepsilon_n\left(\hat{L}|\phi_n\rangle\right)\,,$$

showing that the state $\hat{L}|\phi_n\rangle$ belongs to the same energy level n as $|\phi_n\rangle$. This means that the symmetric operation $\hat{L}$ does not change the energy level structure of $\hat{H}$. This is the formal representation of the fact that "the Hamiltonian is symmetric with respect to the operation $\hat{L}$".

To summarize, when the considered Hamiltonian has the symmetry about the operation $\hat{L}$, it is derived from Eq. (11.35) that

$$\hat{L}\hat{H}=\hat{H}\hat{L}\,,\quad\therefore\quad \hat{L}\hat{H}-\hat{H}\hat{L}=0\,. \tag{11.36}$$

Defining the commutator between two operators as

$$\left[\hat{H},\hat{L}\right] := \hat{H}\hat{L} - \hat{L}\hat{H}\,, \tag{11.37}$$

then the symmetry is represented as

$$\left[\hat{H},\hat{L}\right] = 0\,. \tag{11.38}$$

11.3.2 Energy Index Sorted by Symmetry

Using the irreducible representation, $|\psi_{\Gamma_j}\rangle$, of a symmetric operation,

$$\hat{L}|\psi_{\Gamma_j}\rangle = \Gamma_j \cdot |\psi_{\Gamma_j}\rangle \, ,$$

let us put them from both side of Eq. (11.38) as

$$\langle \psi_{\Gamma_j} | \left[\hat{H}, \hat{L}\right] |\psi_{\Gamma_k}\rangle = 0 \, .$$

Then it is evaluated as

$$\langle \psi_{\Gamma_j} | \left(\hat{H}\hat{L} - \hat{L}\hat{H}\right) |\psi_{\Gamma_k}\rangle = (\Gamma_k - \Gamma_j) \, \langle \psi_{\Gamma_j} |\hat{H}|\psi_{\Gamma_k}\rangle = 0 \, ,$$

to lead,

$$\langle \psi_{\Gamma_j} |\hat{H}|\psi_{\Gamma_k}\rangle = 0 \quad , \quad (\Gamma_j \neq \Gamma_k) \, , \tag{11.39}$$

as the zero off-diagonal property.

Generally, a base $|\psi_{\Gamma_j}\rangle$ would transform as

$$\hat{H}|\psi_{\Gamma_j}\rangle = \sum_k c_{kj} \cdot |\psi_{\Gamma_k}\rangle \, ,$$

hybridizing with other bases. The requirement Eq. (11.39) says, however, that

$$c_{kj} = \langle \psi_{\Gamma_j} |\hat{H}|\psi_{\Gamma_k}\rangle = \delta_{kj} \, , \tag{11.40}$$

namely, no hybridization to lead that

$$\hat{H}|\psi_{\Gamma_j}\rangle = \varepsilon_{\Gamma_j} \cdot |\psi_{\Gamma_j}\rangle \, .$$

It is therefore the base, $|\psi_{\Gamma_j}\rangle$, is the eigenstate originally of $\hat{L}$, but also of Hamiltonian. The energy level can then be labelled by the index of the irreducible representation of $\hat{L}$ as

$$\varepsilon = \varepsilon_{\Gamma_j} \, . \tag{11.41}$$

11.4   Wavevector, Reciprocal Lattice, Brillouin Zone

In this section, we introduce the Brillouin zone as understood as the 'unit circle' for a three dimensional non-orthogonal lattice. In undergraduate courses with limited years, it sometimes occurs that the new topics appear not in the appropriate order (e.g., concepts of quantum physics appear before quantum mechanics is taught). The concept of Brillouin zone may appear in the earlier course of 'semi-conductor devices' etc., where it is introduced with Bloch's theorem (Sect. 11.4.3) for the electronic state specification. In the context, it is shown that the shift in the specification of $\mathbf{k}$ by reciprocal lattices, $\mathbf{k} + \mathbf{G}$, doesn't matter, and then it is sufficient to specify a state within the first Brillouin (this is explained later in this section as well). If this is the first contact by the Brillouin zone to the beginners, it sometimes leads to such a misunderstanding that the Brillouin zone is the concept regarding to the electronic wavefunctions. Then, some beginners are confused when it appears in phonon dispersion as 'where is wavefunction in this story?'... The Brillouin zone is the concept not limited to the Bloch's theorem but more general in treating three dimensional non-orthogonal lattices, as emphasized in this section.

11.4.1 Imagine the Wave Number as a Picture

Whether it is a cycle about time ($t = [0, T]$) or space ($x = [0, L]$), it is a quite natural way to regard the repetition captured as a revolution on the unit circle described by an angle θ as

$$t : [0, T] \quad \rightarrow \quad \theta : [0, 2\pi]$$
$$x : [0, L] \quad \rightarrow \quad \theta : [0, 2\pi] \, . \tag{11.42}$$

The most natural choice for the projection function, $x \rightarrow \theta$ or $t \rightarrow \theta$ is the linear proportion $\theta \propto x$ or $\theta \propto t$ as

$$\theta = k \cdot x = \omega \cdot t \, . \tag{11.43}$$

The proportional coefficients, k or ω should be determined so that $x = L$ or $t = T$ may correspond to $\theta = 2\pi$ when substituted into Eq. (11.43), and hence,

$$k = \frac{2\pi}{L} \quad , \quad \omega = \frac{2\pi}{T} \, . \tag{11.44}$$

Then, let us consider a 1-dim. periodic array of a basic grid a. Taking a period as 'one cycle by advancing the basic grid by N'. In this case, the interval $x = [0, N \times a]$ corresponds to $\theta = [0, 2\pi]$, and hence,

$$k \cdot x = \frac{2\pi}{N \cdot a} x = \frac{2\pi / N}{a} x =: \frac{\varphi}{a} x \, . \tag{11.45}$$

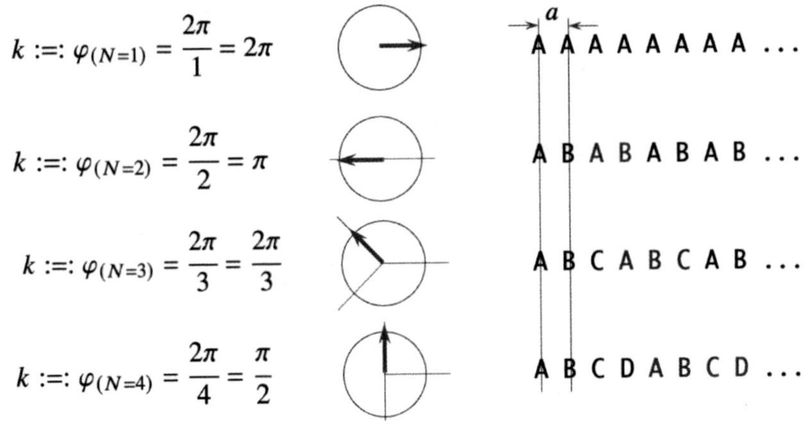

$$k := \varphi_{(N=1)} = \frac{2\pi}{1} = 2\pi$$ A A A A A A A A ...

$$k := \varphi_{(N=2)} = \frac{2\pi}{2} = \pi$$ A B A B A B A B ...

$$k := \varphi_{(N=3)} = \frac{2\pi}{3} = \frac{2\pi}{3}$$ A B C A B C A B ...

$$k := \varphi_{(N=4)} = \frac{2\pi}{4} = \frac{\pi}{2}$$ A B C D A B C D ...

Fig. 11.3 Picturized image of the wavenumber k. The $(N + 1)$-th site gets back to the 1st site in its periodicity, being specified as $k = (2\pi/N)$

The parameter a is system-dependent, and sometimes ignored to be regarded as unity as a convention, leading to the identical treatment

$$k := \varphi_N = \frac{2\pi}{N} . \tag{11.46}$$

This relation can be picturized as in Fig. 11.3. Here, the $(N + 1)$-th site gets back to the 1st site in its periodicity, being specified as $k = (2\pi/N)$.

11.4.2 Introducing Reciprocal Lattice

The area of definition of the unit circle is $\theta = [-\pi, \pi]$ when we keep the symmetrically to the origin. Let's reconsider how the boundary of this domain (zone boundary/ZB) is determined. We can regard it to be determined from

$$|k^{(ZB)}| = \pi = (2\pi)/2 = G/2 ,$$

and, henceforth, we shall consider that 'the bisector of the interval $[0, G]$ specifies the zone boundary'.

By considering the role played by $G = 2\pi$ in a 1-dim. lattice with the unit circle, we extend the concept of **G** working for a 3-dim. non-orthogonal lattice with the similar role. Then, the region surrounded by its bisecting surface is the corresponding extension of the 'unit circle', that is the first Brillouin zone, as explained below.

From Fig. 11.3 and Eq. (11.45), the '2π' in the 1-dim. system actually corresponds to

$$2\pi := \frac{2\pi}{a} . \tag{11.47}$$

The reason why such a quantity appears was originated from the phase factor,

$$\exp(i \cdot \theta) = \exp(i \cdot k \cdot x) = \exp\left(i \cdot \frac{2\pi}{a} \cdot x\right), \tag{11.48}$$

where '$G = 2\pi :=: 2\pi/a$' plays the role to give 'zero phase shift' ($\theta = 2\pi \cdot n :=: 0$) when substituting $x = a \times n$ (n ; integer). Preparing for the concept extension introduced later, let us denote

$$G_0 := \frac{2\pi}{a} \quad , \quad G_m := G_0 \cdot m \quad , \quad R_n = a \cdot n \in \{X\}, \tag{11.49}$$

where G_m is the lattice in k-space formed as the multiples of G_0, and R_n is a point on the grid $\{X\}$ in the real space with the lattice constant a (m, n ; integer). By using the notation, it is expressed

$$\exp(i \cdot G_m \cdot R_n) = \exp(i \cdot 2\pi \cdot n \times m) :=: \exp(i \cdot 0). \tag{11.50}$$

Then, for a given real lattice R_n, we can regard G_0 as *defined* so that it can satisfy

$$\exp(i \cdot G_m \cdot R_n) = \exp[i \cdot 2\pi N] \quad , \quad G_m := G_0 \cdot m. \tag{11.51}$$

For the $G_0 = 2\pi/a$ defined as above, the zone boundary is *determined* as the bisecting surface,

$$|k^{(ZB)}| = G_0/2 \tag{11.52}$$

as we can regard it as a procedure, which will be extended to 3-dim. case. The grid G_m introduced in this context is termed as the **reciprocal lattice** corresponding to the real lattice R_n. The domain inside the $|k^{(ZB)}|$ is the unit circle for 1-dim. system.

Then, let us consider a 3-dim. non-orthogonal lattice with primitive lattice vectors, $\{\mathbf{a}_\alpha\}_{\alpha=1,2,3}$, as

$$\{\mathbf{X}\} \ni \mathbf{R}_n = \sum_{\alpha=1}^{3} n_\alpha \cdot \mathbf{a}_\alpha \quad , \quad n_\alpha \in \mathbb{N}. \tag{11.53}$$

For any translation vector $\mathbf{R}_n$, let us define its reciprocal quantity $\mathbf{G}$ so that

$$\exp[i\mathbf{G} \cdot \mathbf{R}_n] = \exp[i \cdot 2\pi N], \tag{11.54}$$

as we did in 1-dim. case above in Eq. (11.51).

As G_m for 1-dim., let us consider the $\mathbf{G}$ given as the grid in $\mathbf{k}$-space,

$$\mathbf{G_m} \in \left\{ \sum_{\alpha=x,y,z} m_\alpha \cdot \mathbf{b}_\alpha \right\}_{\{m_\alpha\} \in \mathbb{N}},$$

as multiples of primitive vectors $\{\mathbf{b}_\alpha\}_{\alpha=1,2,3}$ in $\mathbf{k}$-space. Then, it leads to

$$\mathbf{G_m} \cdot \mathbf{R_n} = \sum_{\alpha,\beta} m_\alpha \mathbf{b}_\alpha \cdot n_\beta \mathbf{a}_\beta = \sum_{\alpha,\beta} m_\alpha n_\beta \cdot (\mathbf{b}_\alpha \cdot \mathbf{a}_\beta) \ .$$

If $\{\mathbf{b}_\alpha\}_{\alpha=1,2,3}$ are designed to satisfy

$$\mathbf{b}_\alpha \cdot \mathbf{a}_\beta = 2\pi \cdot \delta_{\alpha\beta} \ , \tag{11.55}$$

then

$$\mathbf{G_m} \cdot \mathbf{R_n} = \sum_{\alpha,\beta} m_\alpha n_\beta \left(\mathbf{b}_\alpha \cdot \mathbf{a}_\beta\right) = \sum_{\alpha,\beta} m_\alpha n_\beta \left(2\pi \delta_{\alpha\beta}\right) = 2\pi \sum_\alpha m_\alpha n_\alpha \ ,$$

can surely achieve $\exp[i\mathbf{G_m} \cdot \mathbf{R_n}] = \exp[i \cdot 2\pi N]$ $(N \in \mathbb{N})$ as we want.

Such $\{\mathbf{b}_\alpha\}_{\alpha=1,2,3}$ (primitive vectors to form a reciprocal lattice) to satisfy Eq. (11.55) can be constructed using the property of vector products as follows: The $\delta_{\alpha\beta}$ in Eq. (11.55) means the requirement that $\mathbf{b}_1$ should be orthogonal both to $\mathbf{a}_2$ and $\mathbf{a}_3$. Since the vector product $(\mathbf{a}_2 \times \mathbf{a}_3)$ has the property to be orthogonal both to $\mathbf{a}_2$ and $\mathbf{a}_3$, we can set $\mathbf{b}_1 = A \cdot (\mathbf{a}_2 \times \mathbf{a}_3)$ with a proportional coefficient A. The coefficient can be determined from Eq. (11.55) as

$$\mathbf{a}_1 \cdot \mathbf{b}_1 = A \cdot \mathbf{a}_1 \cdot (\mathbf{a}_2 \times \mathbf{a}_3) \overset{!}{=} 2\pi$$
$$\therefore \ A = \frac{2\pi}{\mathbf{a}_1 \cdot (\mathbf{a}_2 \times \mathbf{a}_3)} \ .$$

As such,

$$\mathbf{b}_1 = 2\pi \cdot \frac{\mathbf{a}_2 \times \mathbf{a}_3}{\mathbf{a}_1 \cdot (\mathbf{a}_2 \times \mathbf{a}_3)} = 2\pi \cdot \frac{\mathbf{a}_2 \times \mathbf{a}_3}{\Omega} \ , \tag{11.56}$$

where $\Omega = \mathbf{a}_1 \cdot (\mathbf{a}_2 \times \mathbf{a}_3)$ is the volume of the unitcell in real space. Doing the same also for other subscripts of $\{\mathbf{b}_\alpha\}_{\alpha=1,2,3}$, the relationship can be summarized using the Eddington ε-symbol [10] as

$$\mathbf{b}_i = \frac{2\pi}{\Omega} \cdot \varepsilon_{ijk} \left(\mathbf{a}_j \times \mathbf{a}_k\right) \ .$$

This is the well-known formula of the reciprocal lattice vectors defined for given primitive lattice vectors in real space, $\{\mathbf{a}_\alpha\}_{\alpha=x,y,z}$, as found in any textbook of Solid State Physics [4]. In this section, we have explained what is the background motivation to lead such a definition.

Let us summarize the procedure: For a 3-dim. non-orthogonal lattice, $\{\mathbf{R_n}\}$ as given in Eq. (11.53), its reciprocal lattice is defined so that

$$\exp[i\mathbf{G_m} \cdot \mathbf{R_n}] = 1 \tag{11.57}$$

is satisfied. For the definition of the zone boundary as in Eq. (11.52) for 1-dim., $\mathbf{G_0}$ in this case corresponds to every single $\{\mathbf{b}_\alpha\}_{\alpha=1,2,3}$ vector, forming a 3-dim. bisecting domain

$$\left[-\frac{\mathbf{G}_0}{2}, \frac{\mathbf{G}_0}{2}\right] = \left[-\frac{\mathbf{b}_\alpha}{2}, \frac{\mathbf{b}_\alpha}{2}\right] \, ,$$

around the origin. This domain is the extended concept of the *unit circle* ($[-\pi, \pi]$) for 3-dim. non-orthogonal lattice, called **the first Brillouin zone** [4].

11.4.3 Bloch Function

As we explained in Sect. 11.2.3, the index (Γ) of the eigenfunction for the symmetry operation $\hat{L}$,

$$\hat{L}\, \bar{\psi}_\Gamma\,(\mathbf{r}) = \bar{\psi}\left(\hat{R} \cdot \mathbf{r}\right) = \lambda_\Gamma \cdot \bar{\psi}_\Gamma\,(\mathbf{r}) \, , \tag{11.58}$$

classifies the energy level.

Considering the translational symmetry for $\hat{L}$,

$$\hat{T}_\mathbf{R} \cdot \psi\,(\mathbf{r}) = \psi\,(\mathbf{r} + \mathbf{R}) \, ,$$

it is shown that the following function,

$$\psi_\mathbf{k}\,(\mathbf{r}) = \exp\left[i \cdot \mathbf{k} \cdot \mathbf{r}\right] \cdot u\,(\mathbf{r}) \quad , \quad u\,(\mathbf{r} + \mathbf{R}) = u\,(\mathbf{r}) \, , \tag{11.59}$$

satisfies the relation as the eigenfunction for $\hat{T}_\mathbf{R}$, namely,

$$\hat{T}_\mathbf{R} \cdot \bar{\psi}_\mathbf{k}\,(\mathbf{r}) = \lambda_\mathbf{k} \cdot \bar{\psi}_\mathbf{k}\,(\mathbf{r}) \, ,$$

which is called **Bloch function** [4]. This relation (**/Bloch's theorem**) is verified as

$$\hat{T}_\mathbf{R} \cdot \left[\exp\left[i \cdot \mathbf{k} \cdot \mathbf{r}\right] \cdot u\,(\mathbf{r})\right] = \exp\left[i \cdot \mathbf{k}\,(\mathbf{r} + \mathbf{R})\right] \cdot u\,(\mathbf{r} + \mathbf{R})$$
$$= \exp\left[i \cdot \mathbf{k} \cdot \mathbf{R}\right] \cdot \exp\left[i \cdot \mathbf{k} \cdot \mathbf{r}\right] \cdot u\,(\mathbf{r})$$
$$\therefore \quad \hat{T}_\mathbf{R} \cdot \bar{\psi}_\mathbf{k}\,(\mathbf{r}) = \exp\left[i \cdot \mathbf{k} \cdot \mathbf{R}\right] \cdot \bar{\psi}_\mathbf{k}\,(\mathbf{r}) \, , \tag{11.60}$$

with the eigenvalue $\lambda_\mathbf{k} = \exp\left[i \cdot \mathbf{k} \cdot \mathbf{R}\right]$ indexed by the wavevector $\mathbf{k}$. The above relation can read that the Bloch function earns the phase factor $\exp\left[i \cdot \mathbf{k} \cdot \mathbf{R}\right]$ when it gets a translation by $\mathbf{R}$.

For the eigenvalue relation with the shift by a reciprocal lattice vector $\mathbf{G}$,

$$\hat{T}_\mathbf{R} \cdot \bar{\psi}_{\mathbf{k}+\mathbf{G}}\,(\mathbf{r}) = \lambda_{\mathbf{k}+\mathbf{G}} \cdot \bar{\psi}_{\mathbf{k}+\mathbf{G}}\,(\mathbf{r}) \, , \tag{11.61}$$

we can derive

$$\lambda_{\mathbf{k}+\mathbf{G}} = \exp\left[i \cdot (\mathbf{k} + \mathbf{G}) \cdot \mathbf{R}\right] = \exp\left[i \cdot \mathbf{k} \cdot \mathbf{R}\right] = \lambda_{\mathbf{k}}, \qquad (11.62)$$

where we used the relation $\exp\left(i\mathbf{G} \cdot \mathbf{R}\right) = 1$ as given in Eq. (11.57). It is then read the fact that the shift by any reciprocal lattice vector $\mathbf{G}$ does not matter to the specification of $\lambda_{\mathbf{k}}$ and therefore to the energy level. It means that the energy level of a translationally symmetric system can be specified within the first Brillouin zone.

11.5 Twisted Boundary Condition

11.5.1 Mesh Shift

One should be aware of the difference between 'the phase gain over an unit cell' and 'the phase gain over a simulation cell' so as not to be confused. The phase gain over an unit cell is specified in Bloch's theorem explained in the previous section

$$\psi_k\left(r + a\right) = \exp\left[ika\right] \cdot \psi_k\left(r\right),$$

via the Bloch function labeled by k. For this Bloch function, the phase gain over a simulation cell (let L times the unit cell) is

$$\psi_k\left(r + L \cdot a\right) = \exp\left[ik\left(L \times a\right)\right] \cdot \psi_k\left(r\right).$$

It is very important to realize that this is merely a statement of a consequence and that there is essentially **no condition to be imposed at this point**. Just in order to simplify the calculation, we would impose the relation between $\psi_k\left(r + L \cdot a\right)$ and $\psi_k\left(r\right)$ as

$$\exp\left[ik\left(L \times a\right)\right] \overset{!}{=} \exp\left[i \cdot \varphi_T\right]. \qquad (11.63)$$

The '!' mark above the equals sign means that the relation is artificially imposed with a fixed value φ_T. φ_T is the phase angle acquired by the Bloch function when it advances by a simulation cell, and the simplest setting for it is

$$\varphi_T = 0,$$

known as **Born-von Karman's periodic boundary condition** [4]. Note again that this setting is **simply one choice** and other choices are also possible (see below).

If we take the Born-von Karman condition, we obtain from Eq. (11.63) that

$$k\left(L \times a\right) = 0 + 2\pi M \qquad (11.64)$$

$$\therefore \quad k = \frac{2\pi}{a} \times \frac{1}{L} \times M, \qquad (11.65)$$

where M is an integer. It then follows that the k index of the Bloch function corresponds to a point on the mesh discretized by the mesh divided by L over a Brillouin zone (of size $2\pi/a$).

On the other hand, there is nothing wrong not to take the Born-von Karman periodic boundary condition but taking more general possibility

$$\varphi_T = \phi_S + 2\pi M ,$$

with non-zero ϕ_S. This corresponds to set the phase angle acquired by the Bloch function to advance by a simulation cell being non-zero ϕ_S, called **twisting boundary condition** [11]. In this case, Eq. (11.65) is replaced by

$$k = \frac{\phi_S + 2\pi M}{a} \times \frac{1}{L} = \frac{\phi_S}{aL} + \frac{2\pi}{a} \times \frac{1}{L} \times M$$
$$= k_S + \frac{2\pi}{a} \times \frac{1}{L} \times M . \qquad (11.66)$$

This leads to the grid to specify the k index of the Bloch function being shifted from the origin by k_S (**shifted k-mesh**).

As explained at the end of Sect. 4.2.1, the calculation with the discretized mesh dividing the Brillouin zone into L segments corresponds to an approximated description within L times of the unitcell. This is an approximation that the component of the wave function with the wavelength more than L times of the unitcell size is treated as negligible. This is what is omitted from the description by treating the integral over the Brillouin zone replaced by the sum over the discretized mesh. Regardless of whether the discretized mesh is aligned with the origin in k-space [Born-von Karman boundary condition ($\phi_S = 0$)] or shifted slightly from it [twisted boundary condition ($\phi_S \neq 0$)], the longer wavelength components cut off in the approximation are the same. One would naively wonder that Born-von Karman boundary condition ($\phi_S = 0$) would give better description by its higher symmetric mesh, but it is known that the finite size error[10] is smaller when a shifted mesh is used [12]. In DFT package, one can specify to use a shifted mesh ($\phi_S \neq 0$) in the input file [13].

11.5.2 Twisting Average

Electrical conductivity can be viewed as a long distance propagation of the influence of the wave function triggered at a some point. The **phase stiffness** [14] of the wave function measures such a property. Let us consider the stiffness against the twisting of the phase angle of a wavefunction between two points separated by a distance. For a conductive system like a metal, the wavefunction retains its influence over long distances, leading to the twist on one side transmitted as a response to the far side, generating a finite stiffness. This corresponds to the image that the phase maintains

---

[10] The error caused by the finite size of the simulation cell is called the finite size error.

coherency over long distances. On the other hand, in an insulator, the coherency is broken in a short distance, so that even if one side is twisted, the effect does not reach far, giving the phase stiffness being zero. Thus, a formalism is possible to link the phase stiffness with the conductivity of the system.[11]

In this view, we can understand a metallic system as a system composed of relevant components with longer wavelength. In metallic systems, therefore, we have to handle the truncation of long wavelength modes (small k components) more carefully. A discretized grid with a mesh width of Δk cannot describe longer wavelength modes in the $k = [0, \Delta k]$ interval. Since metallic systems are generally serious on long wavelength modes, the missing $[0, \Delta k]$ can be a significant error factor. To reduce this error, the meshshift k_S in Eq. (11.66) is randomly selected and the calculations with the various twisted boundary conditions are repeated and averaged (**twisting averaging** [11,15]).

In the reciprocal space picture, the twisting averaging can be understood more directly. As described in Sect. 4.2.3, the Fermi surface for a metallic system takes more complex shape, which gets to be difficult to be captured by a coarse k mesh. Therefore, the idea is to make efforts to capture it by repeating various evaluations with shifted k-mesh with different amounts of the shift k_S and averaging over the evaluations.

Twisting averaging can be understood from a further different perspective. Periodic boundary conditions fixed to some ϕ_S, such as the Born-von Karman condition, are only an artificial convenience. Nature should not feel this, but in actual calculations, a dependence on the twisting phase angle ϕ_S arises in the results. This is because the originally infinite system was mocked up with L-fold extended cell iterations. We have created an artificial wall at the L-fold extension to fix the phase angle ϕ_S there, but in reality, no such wall exists. The reason why the predication depends on the choice of ϕ_S is that the calculation unwillingly *feels* the existence of such an artificial wall. By taking the twisting averaging, such artificial dependence that should not be there originally is cancelled out to get the true value.

11.5.3   A Rough Description of Polarization Theory

The polarization P in materials follows the relation with the current j [9] as

$$\mathbf{j} = \frac{d\mathbf{P}}{dt} \ , \quad \mathbf{P} = e \cdot \mathbf{x} . \tag{11.67}$$

Via this relation, the current relates with the mean value of the position $\hat{\mathbf{x}}$,

$$\langle \mathbf{j} \rangle \sim \langle \mathbf{x} \rangle = \langle \psi_{\mathbf{k}} | \hat{\mathbf{x}} | \psi_{\mathbf{k}} \rangle . \tag{11.68}$$

---

[11] The topic leads to the Kohn's concept of 'near-sightedness'. More advanced formulation has been developed by King-Smith and Vanderbilt, leading to the polarization theory formulated in terms of the Berry phase (Sect. 11.5.3).

For a contribution with the phase angle $\mathbf{k}$ to a wave function,

$$\psi_{\mathbf{k}} \sim \exp\left(i\,\mathbf{k}\cdot\mathbf{x}\right),\tag{11.69}$$

the position operator is expressed as

$$\hat{\mathbf{x}}|\psi_{\mathbf{k}}\rangle \sim -i\,\boldsymbol{\nabla}_{\mathbf{k}}|\psi_{\mathbf{k}}\rangle.\tag{11.70}$$

As such, the current relates with the mean value

$$\langle \mathbf{j}\rangle \sim -i\,\langle \psi_{\mathbf{k}}|\boldsymbol{\nabla}_{\mathbf{k}}|\psi_{\mathbf{k}}\rangle.\tag{11.71}$$

The quantity on the right-hand side appearing here is related to what called **Berry phase** or **connection** in geometry [14] and has the following meaning. When a field quantity changes in a space characterized by the parameter λ, we can expand the infinitesimal change with respect to λ as

$$|\psi_{\lambda+\Delta\lambda}\rangle = |\psi_{\lambda}\rangle + \Delta\lambda\cdot\boldsymbol{\nabla}_{\lambda}|\psi_{\lambda}\rangle + \cdots,\tag{11.72}$$

leading to

$$\delta\langle\psi_{\lambda}|\psi_{\lambda+\Delta\lambda}\rangle \sim \Delta\lambda\cdot\langle\psi_{\lambda}|\boldsymbol{\nabla}_{\lambda}|\psi_{\lambda}\rangle.\tag{11.73}$$

Taking $\lambda = \mathbf{k}$, Eq. (11.71) is then

$$\langle\mathbf{j}\rangle \sim \delta\langle\psi_{\lambda}|\psi_{\lambda+\Delta\lambda}\rangle,\tag{11.74}$$

where the quantity $\langle\psi_{\lambda}|\psi_{\lambda+\Delta\lambda}\rangle$ means the change of the field induced by the parameter change $\psi_{\lambda+\Delta\lambda}$, measured at the reference point $\langle\psi_{\lambda}|$.

Here, we can see the interpretation of $\langle\mathbf{j}\rangle \sim \delta\langle\psi_{\mathbf{k}}|\psi_{\mathbf{k}+\Delta\mathbf{k}}\rangle$ as understood that the difference between metals and insulators intuitively characterized by $\mathbf{j}$ is related to the stiffness of the phase angle of the wave function. This consequence can fully be derived [16] in the framework of quantum mechanical linear response theory applied to the susceptibility σ in the relation

$$\mathbf{j} = \sigma\cdot\mathbf{E}.\tag{11.75}$$

References

1. Condensed Matter in a Nutshell Gerald D. Mahan, Princeton Univ Press (2010/10/4) ISBN-13:978-0691140162
2. "Electron Correlation in the Solid State", N.H. March (ed.), Imperial College Press (1999), ISBN:1-86094-200-8

3. "Principles of the Theory of Solids", J. M. Ziman, Cambridge University Press (1972/7/20) ISBN-13:978-0521083829
4. "Solid State Physics", N.W. Ashcroft and N.D. Mermin, Thomson Learning (1976).
5. "Semiconductor Devices, Physics and Technology" Simon Sze and Ming-Kwei Lee, Wiley (2015/1/1) ISBN-13:978-8126556755
6. "Density Functional Theory: A Practical Introduction", David S. Sholl, Janice A. Steckel, Wiley-Interscience (2009/4/13), ISBN-13:978-0470373170.
7. "Adhesion of Electrodes on Diamond (111) Surface: A DFT Study", T. Ichibha, K. Hongo, I. Motochi, N.W. Makau, G.O. Amolo, R. Maezono, Diam. Relat. Mater. 81, 168 (2018). https://doi.org/10.1016/j.diamond.2017.12.008
8. "Physical Chemistry: A Molecular Approach" Donald A. McQuarrie, John D. Simon, University Science Books (1997/8/20) ISBN-13:978-0935702996
9. "Classical Electrodynamics", John David Jackson, Wiley (1998/8/14), ISBN-13:978-0471309321
10. "Mathematical Methods for Physics and Engineering" Ken F. Riley, Mike P. Hobson, Stephen J. Bence, (Third Edition), Cambridge University Press (2006/6/8) ISBN-13:978-0521683395
11. "Twist-averaged boundary conditions in continuum quantum Monte Carlo algorithms", C. Lin, F.H. Zong, and D.M. Ceperley, Phys. Rev. E 64, 016702 (2001). https://link.aps.org/doi/10.1103/PhysRevE.64.016702
12. "Mean-Value Point in the Brillouin Zone", A. Baldereschi, Phys. Rev. B 7, 5212 (1973). https://link.aps.org/doi/10.1103/PhysRevB.7.5212
13. https://www.quantum-espresso.org (URL confirmed on 2022.11)
14. "Electronic Structure: Basic Theory and Practical Methods", Richard M. Martin, Cambridge University Press (2004/4/8), ISBN-13:978-0521782852
15. "Diamond to beta-tin phase transition in Si within quantum Monte Carlo" R. Maezono, N.D. Drummond, A. Ma, and R.J. Needs, Phys. Rev. B 82, 184108 (2010). https://doi.org/10.1103/PhysRevB.82.184108
16. "Berry Phases in Electronic Structure Theory: Electric Polarization, Orbital Magnetization and Topological Insulators", David Vanderbilt, Cambridge University Press (2018/11/1), ISBN-13:978-1107157651

Appendix D: A Brief Explanation of DFT+U

12

Abstract

This section summarizes topics related to many-body electron theory that is cited in connection with the exchange–correlation potentials in the main text. Sections 12.1–12.2 are cited from the main text at Sect. 5.3.7. Section 12.3 is cited from the main text at Sect. 5.3.3.

12.1 The Problem of Damage to Self-interaction Cancellation

12.1.1 Cancellation of Self-interaction

The exact solution of the many-body Schrodinger equation (5.1) minimizes the value of the integral

$$\frac{\int d\mathbf{r}_1 \cdots \mathbf{r}_N \cdot \Psi^* (\mathbf{r}_1 \cdots \mathbf{r}_N) \cdot \hat{H} \Psi (\mathbf{r}_1 \cdots \mathbf{r}_N)}{\int d\mathbf{r}_1 \cdots \mathbf{r}_N \cdot \Psi^* (\mathbf{r}_1 \cdots \mathbf{r}_N) \Psi (\mathbf{r}_1 \cdots \mathbf{r}_N)} = I , \qquad (12.1)$$

known as **variational principle** [1]. Substituting Assumption (5.4) for the functional form of the many-body wave function as the Slater determinant into the above, the integral becomes

$$I = I[\{\psi_j\}_1^N] , \qquad (12.2)$$

as the integral value depending on N-fold orbital functions $\{\psi_j\}$ (variational functional with respect to the orbital functions).

© The Author(s), under exclusive license to Springer Nature Singapore Pte Ltd. 2023
R. Maezono, *Ab initio Calculation Tutorial*,
https://doi.org/10.1007/978-981-99-0919-3_12

The "solution" to minimize this functional is obtained in terms of "the equation to be satisfied by $\{\psi_j\}$"[1] as

$$
\left[-\frac{1}{2}\nabla^2 + v_{ext}\left(\mathbf{r}\right) + \sum_{j \neq i} \int d\mathbf{r}' \frac{\left|\psi_j\left(\mathbf{r}'\right)\right|^2}{\left|\mathbf{r} - \mathbf{r}'\right|} \right] \cdot \psi_i\left(\mathbf{r}\right) \tag{12.3}
$$

$$
- \sum_{j \neq i} \left[\int d\mathbf{r}' \psi_j^*\left(\mathbf{r}'\right) \frac{\delta_{\sigma_i \sigma_j}}{\left|\mathbf{r} - \mathbf{r}'\right|} \psi_i\left(\mathbf{r}'\right) \right] \cdot \psi_j\left(\mathbf{r}\right)
$$

$$
= \varepsilon_i \cdot \psi_i\left(\mathbf{r}\right) .
$$

This is the Hartree–Fock equation which appeared in Sect. 5.3.1 [3]. Among four terms appeared in the left-hand side, the third (fourth) term is called Hartree (Fock) term.

The Slater determinant (5.4) is an approximate assumption for the many-body wave function, which is an upgraded version of Eq. (5.2) which is the simplest variable separation. The upgrade takes into account the quantum mechanical exchange effect (e.g., Pauli exclusion rule). An approximation to perform the variational minimization of Eq. (12.1) using Assumption (5.2) is called the Hartree approximation [3]. On the other hand, the approximation using the upgraded Assumption (5.4) is called the Hartree–Fock approximation to get the Hartree–Fock equation. The Hartree term in Eq. (12.4) commonly appears in both approximations, but the Fock term is the one that does not appear in the Hartree approximation. The difference between the two is whether or not the Pauli exchange is taken into account, thus the Fock term expresses the quantum exchange effect.

For the third and fourth terms on the left-hand side of the Hartree–Fock equation, we rewrite them as follows to better interpret the meaning of these interactions,

$$
V^{(HF)} = \sum_{j \neq i} V_{ij}^{(H)} + \sum_{j \neq i} V_{ij}^{(X)} = V^{(H)} + V^{(X)} , \tag{12.4}
$$

$$
V_{ij}^{(H)} := \int d\mathbf{r}' \frac{\left|\psi_j\left(\mathbf{r}'\right)\right|^2}{\left|\mathbf{r} - \mathbf{r}'\right|} \cdot \psi_i\left(\mathbf{r}\right) = \int d\mathbf{r}' \frac{n_j\left(\mathbf{r}'\right)}{\left|\mathbf{r} - \mathbf{r}'\right|} \cdot \psi_i\left(\mathbf{r}\right) , \tag{12.5}
$$

$$
V_{ij}^{(X)} := - \int d\mathbf{r}' \psi_j^*\left(\mathbf{r}'\right) \frac{\delta_{\sigma_i \sigma_j}}{\left|\mathbf{r} - \mathbf{r}'\right|} \psi_i\left(\mathbf{r}'\right) \cdot \psi_j\left(\mathbf{r}\right) , \tag{12.6}
$$

where the superscripts mean "HF = Hartree–Fock", "H = Hartree", "X = eXchange". The $n_j\left(\mathbf{r}\right) = \left|\psi_j\left(\mathbf{r}\right)\right|^2$ is the quantity interpreted as the charge density formed by the j-th orbital.

The regulation "$i \neq j$" expresses that interactions occur with other particles, but from the perspective of mathematical handling, these exclusion conditions are like a "a pain in the butt".

---

[1] It is called Euler–Lagrange equation [2] in general context.

In fact, the Hartree term,

$$V^{(H)} = \sum_{j \neq i} V_{ij}^{(H)} ,$$ (12.7)

is taking the form like "almost there" because, if without the restriction, it can be fabricated as

$$\tilde{V}^{(H)} := \sum_{i,j} V_{ij}^{(H)} = \sum_i \int d\mathbf{r}' \frac{1}{|\mathbf{r} - \mathbf{r}'|} \sum_j n_j (\mathbf{r}') \cdot \psi_i (\mathbf{r})$$

$$= \sum_i \int d\mathbf{r}' \frac{n (\mathbf{r}')}{|\mathbf{r} - \mathbf{r}'|} \cdot \psi_i (\mathbf{r}) = \tilde{V}^{(H)} [n] ,$$ (12.8)

being an functional with respect to the charge density $n (\mathbf{r}) = \sum_j n_j (\mathbf{r})$.

As such, letting

$$V^{(H,X)} = \tilde{V}^{(H,X)} - V_{ii}^{(H,X)} ,$$ (12.9)

we can rewrite

$$V^{(HF)} = \tilde{V}^{(H)} [n] + \tilde{V}^{(X)} - \left(V_{ii}^{(H)} + V_{ii}^{(X)} \right)$$

$$= \tilde{V}^{(H)} [n] + \tilde{V}^{(X)} - V^{(SI)} ,$$ (12.10)

where

$$V^{(SI)} := V_{ii}^{(H)} + V_{ii}^{(X)} ,$$ (12.11)

called **self-interaction** ($i = j$, interactions with oneself). From Eqs. (12.5) and (12.5), it leads to $V_{ii}^{(H)} = - V_{ii}^{(X)}$ *within* the Hartree–Fock theory to get

$$V^{(SI)} = 0 \quad \text{(for Hartree-Fock)} ,$$ (12.12)

called "the cancellation of the self-interaction" [4].

12.1.2 Damage to Self-interaction Cancellation

It should be noted that self-interaction is a quantity that *does not originally exist* but is introduced for convenience in order to create $\tilde{V}^{(H,X)}$ excluding the regulation in the summation, "$i \neq j$". The self-interaction is introduced because it does not matter being zero due to the cancellation though it does not originally exist.

As explained in Sect. 5.3.2, DFT is based on representations by n. The $\tilde{V}^{(H)} [n]$ takes a very compatible form with DFT in this sense. On the other hand, $\tilde{V}^{(X)}$ is in the form of a nonlocal integral[2] that cannot be simply expressed by n ($\mathbf{r}$). The self-interaction cancellation $V^{(SI)} = 0$ is guaranteed only when the nonlocal integral is strictly executed.

The nonlocal integral of $\tilde{V}^{(X)}$ is, however, computationally cumbersome and costly, so an approximate strategy is taken to determine it locally just depending on the local density n ($\mathbf{r}$) (**LDA; local density approximation** [4]). In this case, the exchange part $V^{(X)}$ changes from the exact one, and the self-interaction cancellation with $V^{(H)}$ is damaged (Eq. (12.13)). As a result, the self-interactions do not cancel each other out, and effects that do not originally exist are introduced as contaminating biases. This is the problem of **damage to self-interaction cancellation**. A well-known example of the effect of bias contamination is **the bandgap problem**, as discussed later.

To summarize up to here,

- **Hartree–Fock**: It doesn't matter if you add the self-interactions for convenience that don't originally exist, because they cancel out each other.
- **Conventional-DFT**: Since the self-interactions do not cancel out exactly, self-interactions are spuriously introduced to give contaminating bias.

"conventional-DFT" means the DFT using exchange–correlation potentials, which is based on the oldest LDA and its upgraded families.

It is important to note that DFT itself is strictly expressed in n ($\mathbf{r}$). It is certainly true that "the self-interaction cancellation is damaged because *something* is approximately expressed in n ($\mathbf{r}$)" in the above explanation. But if you omit the details and comprehend only the consequences in a cursory manner, it may lead to such misunderstanding that "DFT is an approximation based on the expression only by n ($\mathbf{r}$) (wrong statement)". Again note that

- **DFT** (exact): It is strictly guaranteed that the final E_G [n] can be expressed as a *functional* of n ($\mathbf{r}$),
- **LDA**: An approximation is taken so that the intermediate quantity V_{XC} is approximately expressed as the *function* of n ($\mathbf{r}$) itself.

One should also note the difference between function and functional.

Reminding the notations introduced in the latter part of Sect. 5.3.2, "XC: exchange–correlation", "X: exchange (effects due to the quantum statistical property)", "C: correlation" (effects from the deformation of wave function due to interactions), we can summarize as

---

[2] Its value is not determined locally but after integrating over the entire region.

$$V^{(HF)} = V^{(H)} + V^{(X)} \text{(non-local)} ,$$

$$V^{(LDA)} = \left[V^{(H)} + V^{(X)} \text{(local)}\right] + V^{(C)} \text{(local)} ,$$

$$V^{(\text{exactKS})} = V^{(H)} + V^{(XC)} . \tag{12.13}$$

In $V^{(HF)}$, only the exchange effect is considered, not the correlation effect ($V^{(C)} = 0$) since it adopts a fixed form for the wave function that is justified without interaction. In $V^{(LDA)}$, both the exchange effect $V^{(X)}$ and the correlation effect $V^{(C)}$ are considered to some extent within an approximation.[3] In $V^{(LDA)}$, the approximation to localize the exchange term resulted in damage to self-interaction cancellation in the bracket $[\cdots]$. Note that, in a rigorous DFT operation (assuming it is possible), V_{XC} should exactly be able to realize self-interaction cancellation with $V^{(H)}$ since the overall rigor is guaranteed.[4]

Note also that we wrote "exactKS" in the superscript instead of "exactDFT". The Hartree term appears in the practical DFT because of the Kohn–Sham formalism. In the formalism, we can take advantage in the prospect by separating out $\tilde{V}^{(H)}[n]$ (interpreted as the classical contribution) from the quantum mechanical particle interactions, and just concentrate on the role of V_{XC}. Note that the rigor of DFT is still kept in the Kohn–Sham formalism without losing generality.

12.2 DFT+U Method

12.2.1 Problem of Bandgap Underestimation

The underestimation of the bandgap by DFT has been a well-known problem, and at one time it was actively campaigned as a fatal weakness that questioned the generality of DFT.[5] Upon the detailed clarifications of the mechanism of the gap underestimation problem, several methods have been developed to resolve the problem, one of which is the DFT+U method, which is most well known even among non-expert users.[6] The U parameter appearing there is called "Hubbard's U", and is often explained as being related to the U parameter of the Hubbard model in the theory of magnetism [6]. The U-parameter of the Hubbard model is the intensity of onsite

---

[3] When the author was a graduate student, it is often heard misleading explanations that electronic correlation was *ignored* in LDA (wrong statement).

[4] It is known that exchange and correlation effects cooperatively work to realize conservation laws, etc. that must be satisfied as a whole. It is a matter of convenience to treat V_X and V_C separately. They shall be treated basically as V_{XC} in a quantity.

[5] However, many of them were sensationalistic in their attempts to emphasize electronic correlation as a catchy keyword. Many of them were not based on sufficient understanding of the mechanism of the exchange effect as described in this section.

[6] Not only the DFT+U, there are several other methods to cope with the problem, such as SIC method (self-interaction correction), hybrid functional methods, etc. [5], those are the dealing at the exchange–correlation functional level. The GW method [5] is another representative method, but it is beyond the level to tune exchange–correlation functionals.

repulsion that explains anti-ferromagnetism upon the superexchange interactions [6]. Considering a model of electrons hopping among the atomic sites, we introduce the model interaction such that the energy increases by U when a same site is occupied by two electrons. Such a double occupancy is possible only for antiparallel spin pairs because the parallel spin pairs are inherently impossible to occupy the same position due to Pauli exclusion. The double occupancy only possible for antiparallel pairs provides a virtual intermediate state with energy increased by U, leading to an energy gain due to perturbation through the virtual state. The stabilization is explained as the origin of the emergence of anti-ferromagnetism.

The mechanism described by the Hubbard model is so clear and impressive that it seems to be a tragedy that it has led to misunderstandings about the DFT+U method. The $U = 0$ in the Hubbard model describes a situation in which there is no electron–electron interaction. Influenced by this image, some of researchers tend to get such misunderstanding that "the DFT ($U = 0$ in DFT+U) does not originally contain electron–electron interaction" → "DFT does not contain electron correlation"..., leading to the misconception that "electron correlations were forcibly included as a U parameter in DFT". As mentioned so far, DFT is a framework that surely takes into account electron interactions, and the correlation effects (electron correlations) are never neglected, even those are not perfect in the practically used exchange–correlation functionals. As explained below, the U parameter of DFT+U is a correcting parameter for exchange effects rather than correlation effects.

12.2.2 Mechanism of Gap Underestimation

"The excitation by U in the Hubbard model" and the easy-going impression that "the bandgap is underestimated because U is not enough" would often mislead the beginners to misunderstand that the conduction band comes *down* due to the lack of U', leading to the underestimation of the gap. This is not true. As described below, band underestimation can be understood in terms of damage to self-interaction cancellation, where the valence band is pushed *up* to lead the gap underestimation, which is a more accurate picture of the mechanism.

The electrons in the valence band should originally receive interactions from other $(N - 1)$-electrons, but the approximated formulation that counts self-interactions results in interactions from other N-electrons instead of $(N - 1)$, resulting in an overcounting of repulsive interactions. As a result, the energy levels of the valence band are improperly evaluated higher than the reality, pushing upward the valence band. On the other hand, the energy levels comprising the conduction band should be evaluated in the $(N + 1)$-electron system (original N-electrons plus one additional electron). In the approximation where the self-interaction cancellation is damaged, this situation is also evaluated with "interactions from other N-electrons". Since one electron in a $(N + 1)$-electron system should be interacted by other N-electrons, it is quantitatively correct. As such, there is no energy level lift for the conduction band. As a result, the gap between the elevated valence band and the un-lifted conduction band gets narrowed, resulting in the underestimation of the gap [7].

12.2.3 Cancellation of Self-interaction Depending on the Locality

The above mechanism is common to both localized and itinerant systems, but as described below, the narrowing of the gap is more pronounced in localized systems because of the difference in the way the N orbital levels are affected [5]. In an itinerant system, all orbitals overlap each other, so the effect of damage to self-interaction cancellation is shared by N orbitals, resulting in a situation where there are many levels that receive a small amount of lift, which is diluted to $1/N$. On the other hand, in a localized system, N orbitals are not overlapped, and hence the damage is not shared by N orbitals but the several orbitals in the vicinity. These affected orbitals lift their level seriously while other orbitals are not affected. Since a band is a superposition of levels, each small lift of all the orbitals in metallic systems is not additive and the overall lift of the valence band remains small. On the other hand, in a localized system, the affected orbitals with larger lift push up the upper edge of the band, making the gap narrower.

As such, the damage to self-interaction cancellation narrows the gap particularly in the localized system. The damage to self-interaction cancellation is due to the approximated evaluation of the exchange term $V^{(X)}$ as mentioned above (Eq. (12.13)). Thus, the problem of gap underestimation can be attributed to exchange effects rather than correlation effects.

The above difference in localized/itinerant systems can be explained in more direct mathematical terms as follows. The extent of damage to self-interaction cancellation depends on how is the spatial localization of $\psi(\mathbf{r})$ through the integral form of Eqs. (12.5) and (12.6). If we adopt the planewave spread over the whole space for the functional form of $\psi(\mathbf{r})$ to model the itinerant electronic state, the factor $1/k^2$ appears as the Fourier transform of $1/r$, resulting in the factor $1/N$ factorized as a common factor, which vanishes as the number of particles gets to $N \rightarrow \infty$, being negligible contribution of the bias due to the damage of the cancellation. On the other hand, if we adopt the Gaussian function for $\psi(\mathbf{r})$ to model the localized electronic state, the factor $1/N$ cannot be a common factor, and the bias gets to be more pronounced than in an itinerant system.

12.2.4 Strategies to Cope with Gap Underestimation

One possible strategy to eliminate the pathology of gap underestimation is to apply a trick such that when the contribution to the self-interaction ($j = i$) that causes the bias, an adjustment is applied to the potential. This corresponds to DFT+U, and the U parameter adjusts the weight of the adjustment. This is explained again in the next section. Another, more fundamental strategy is known as the bf exact exchange method [5]. This method goes back to Eq. (12.13) and seriously evaluates the nonlocal integrals of $\tilde{V}^{(X)}$ to retain the self-interaction cancellation, just as in $\tilde{V}^{(HF)}$. The way to do this is to stop using "$V^{(XC)}$ expressed in the local $n(\mathbf{r})$" as in LDA and evaluate the integral of Eq. (12.6) for $V^{(X)}$. In this case, the exchange part is not expressed in $n(\mathbf{r})$, but explicitly in the orbit function, so it is also called the

orbital-dependent exchange method. The exact exchange method seems not very suitable for practical use due to the high cost of integral evaluation, just within the scope of basic research.

12.2.5 Strategy in DFT+U

The problem of the gap underestimation originates from the self-interaction contribution ($j = i$), which does not exist originally, but is included for convenience in Eqs. (12.4)–(12.11). The idea of the DFT+U method is to use a projection operator that only flags the $j = i$ contribution in the summation, and to adjust the parameter U for the contribution to eliminate the deficiency due to the damage of the cancellation of the self-interaction.

This projection operator takes the form of "when i and j occupy the same location, the potential is increased by U", which is indeed a design concept similar to the "Hubbard U" setup when the double occupancy occurs in the Hubbard model. We should note, however, the different role of U term between the Hubbard model and DFT+U. In the former, it is "(the evaluation without interaction)+(Hubbard U term)", while in the latter the U term plays the correction applied to "(the evaluation taking interactions into account)". Presumably due to the strong impression of Hubbard model, beginners tend to be confused on this point. Also, $j = i$ does not mean "the electrons are hopping and such a situation occurs", but rather "the $j = i$ contribution in the summation".

Since U parameter is an adjustment parameter, users should basically provide it with a specified value in the input file. Then the question arises how to determine this value. Fundamental policy is to adjust the U value so that the deviated quantity affected by the damage of the self-interaction cancellation may be recovered to the proper value. The most representative target quantity to be affected is the bandgap, so the easiest idea to determine U is to adjust U so that the bandgap may coincide with the experimental value. However, there are such cases sometimes where the adjusted U based on the bandgap coincidence deviates the lattice constants from proper values [8]. As such, there are various arguments in basic research on many-body electron theory on which quantity to be taken as the target for calibrating U [5].

What value to adopt for the U parameter is indeed a difficult question. At the level of this book, I would recommend to take inherit values used in previous studies as a safer choice. Recently, utilities to determine U parameters have become to be included in ab initio packages. Nevertheless, the problem faced in practice is a vexing one as follows. On the one hand, we can think that U would be different in each materials, so "U in a compound" should be different from "U in a pristine material". However, on the other hand, we would face a question when we evaluate energy differences such as ΔH whether we should adopt different U or same U for a compound and each pristine elements (ΔH describes the difference between them) [9]. Permitted to adjust different U individually for a compound and pristine elements, we cannot be quite confident to guarantee the legitimacy of the energy difference for predictions, as we notice.

12.2.6 Summary/Possible Misleadings on DFT+U

The problem of "damage to self-interaction cancellation" is often cited as "self-interaction problem". This is quite incorrect wording which leads to the various misunderstandings, as mentioned in the footnote in the main text (Sect. 5.3.7). This wording might mislead the beginners that there *exists* the self-interaction which is not correctly represented by the theory. Once again, we would like to confirm that the self-interaction *does not originally exist* but was introduced for the sake of convenience. What is not correctly represented is not the self-interaction but the *cancellation* of the self-interaction.

Note that $V^{(LDA)}$ includes $V^{(C)}$ in Eq. (12.13). The underestimation of the gap is caused by the fact that the exchange and correlation are both approximated in such a way that they can be conveniently expressed by n ($\mathbf{r}$), which damages the correct exchange. One should be careful not to be confused by such misleading arguments often found like "the gap underestimation arises because DFT (LDA) has $V^{(C)} = 0$, so U is used to represent the correlation as in the Hubbard model".

12.3   Universal Consequences Derived From Bare Interaction

The usual approach in materials science is to categorize the subject by bonding mechanism, such as covalent bonding, intermolecular bonding, and metallic bonding, and to understand the common characteristics and corresponding properties within each category. Covalent bonding is related to phenomena on the order of tens of thousands of degrees, while intermolecular bonding is related to phenomena on the order of hundreds of degrees. On the other hand, there is an understanding from more broader viewpoint to discuss common characteristics, categorized by the ranks of Coulomb interactions, nuclear interactions, gravitational interactions, etc. Whether it is a covalent bond or an intermolecular bond, they commonly originate from the Coulomb interaction of $1/r$. The **effective interaction** for the degree of freedom of interest may look like a $\sim 1/r^6$ van der Waals bonding or $\sim e^{-ar}/r$ Coulomb screening, but originally they all commonly have a $1/r$-type interaction between charges, called **bare** interaction.

From the mere fact that the interaction is of type $1/r$, it follows as a consequence of spatial scaling[7] that **the virial theorem** [10]. For the kinetic energy $\langle K \rangle$ and the potential energy $\langle U \rangle$ ($\langle \cdots \rangle$ means the average value), the relation $\langle K \rangle = -(1/2) \langle U \rangle$ is obtained as the required consequence. This relation should always be satisfied, regardless of whether the target is composed of a covalent bonding, an intermolecular bonding, hydrogen bonding, etc. Considering Hund's first rule stating that spins take the same direction as much as possible in the ground state of an isolated atom, there are some textbooks explaining the reason of the rule as "spins can avoid collisions

---

[7] The conclusion follows from the argument that if we multiply the length scale by a, the interaction is $1/a$ times larger and the kinetic energy of $\sim \nabla^2$ is scaled as $1/a^2$, etc.

and potential loss by setting spins in parallel pairs". It is actually contrary to the virial theorem [11].[8]

DFT described in this book is a universal framework being valid for all targets governed by bare $1/r$-type Coulomb interactions, regardless of the bonding mechanism. It is said to be inspired [13] by the consequences of the **cusp theorem** [12], which is universally derived from the bare interaction types. The charge density at the nucleus position has a cusp,[9] and the slope of the intercept reflects the atomic number Z of the nucleus. This functional form follows universally only from the fact that the bare interaction is $1/r$ [12].

The geometry of the nuclear positions determines the ground state, which in turn determines the charge density. The cusp theorem states that when looking at the charge density, nuclei exist at the positions where the cusp exists, and the cusp intercept value tells us which element with which atomic numbers is located at each nucleus position. In other words, the most essential information (geometry) of the ground state would be expressed only by the information of the charge density. If this is the case, then it would be possible to completely determine the ground state using the charge density as the fundamental quantity. DFT is said to be based on this insight [13]. If the bare interaction is of a different kind, such as nuclear interactions, the functional form corresponding to the above cusp theorem will also be different, and correspondingly there is another DFT for nuclear systems [14].

References

1. "Density-functional Theory of Atoms And Molecules" (International Series of Monographs on Chemistry), Robert G. Parr, Yang Weitao, Oxford University Press USA (1989/1/1), ISBN-13:978-0195092769
2. "Mathematical Methods for Physics and Engineering" Ken F. Riley, Mike P. Hobson, Stephen J. Bence, (Third Edition), Cambridge University Press (2006/6/8) ISBN-13:978-0521683395
3. "Fundamentals of Condensed Matter Physics", Marvin L. Cohen, Steven G. Louie, Cambridge University Press (2016/5/26) ISBN-13:978-0521513319
4. "Density Functional Theory: A Practical Introduction", David S. Sholl, Janice A. Steckel, Wiley-Interscience (2009/4/13), ISBN-13:978-0470373170.
5. "Strong Coulomb Correlations in Electronic Structure Calculations", Vladimir I Anisimov (ed.), CRC Press (2000/5/1) ISBN-13:978-9056991319
6. "Lecture Notes on Electron Correlation and Magnetism", Patrik Fazekas, World Scientific (1999/5/1) ISBN-13:978-9810224745
7. "Electronic Structure: Basic Theory and Practical Methods", Richard M. Martin, Cambridge University Press (2004/4/8), ISBN-13:978-0521782852

---

[8] Books composed by acquainted authors in many-body theory are indeed carefully written avoiding to take such explanations.

[9] This is always true for real charge densities, but for charge densities obtained in simulations, it is only true for all-electron calculations, not for calculations using pseudopotentials as in the example of this book. This is because in pseudopotential calculations, the potential from the nucleus is not of the $1/r$, but replaced by a smooth model potential.

8. "Bandgap reduction of photocatalytic TiO_2 nanotube by Cu doping", S.K. Gharaei, M. Abbasnejad, R. Maezono, Sci. Rep. 8, 14192 (2018). https://doi.org/10.1038/s41598-018-32130-w

9. "Ab initio thermodynamic properties of certain compounds in Nd-Fe-B system", A.T. Hanindriyo, S. Sridar, K.C. Hari Kumar, K. Hongo, R. Maezono, Comp. Mater. Sci. 180, 109696 (2020). https://doi.org/10.1016/j.commatsci.2020.109696

10. "Quantum Theory of the Electron Liquid", Gabriele Giuliani, Giovanni Vignale, Cambridge University Press (2005/3/31) ISBN-13:978-0521821124

11. "Interpretation of Hund's multiplicity rule for the carbon atom", K. Hongo, R. Maezono, Y. Kawazoe, H. Yasuhara, M.D. Towler, and R.J. Needs, J. Chem. Phys. 121, 7144 (2004). https://doi.org/10.1063/1.1795151

12. "Molecular Electronic-Structure Theory", Trygve Helgaker, Poul Jorgensen, Jeppe Olsen, Wiley (2013/2/18), ISBN-13:978-1118531471

13. "Structural Chemistry and Molecular Biology", Alexander Rich and Norman Davidson (ed.), W.H.Freeman & Co Ltd (1968/6/1) ISBN-13:978-0716701354

14. "Intrinsic-density functionals", J. Engel, Phys. Rev. C 75, 014306 (2007). https://link.aps.org/doi/10.1103/PhysRevC.75.014306

Appendix E: Appendix for Data Scientic Topics

13

Abstract

This chapter provides additional explanations on data science topics in the main text. Section 13.1 is cited from the main text Sect. 7.2.3. Section 13.2 is cited from Sect. 7.3.3.

13.1 Bayesian Update of the Parameters of Distribution

Consider the problem to estimate the parameters $\vec{\theta}$, which dominates the distribution $p\left(z \mid \vec{\theta}\right)$ with respect to the variable z.[1] The key concept of the Bayesian estimation is to estimate $\vec{\theta}$ as a distribution of certainty $p\left(\vec{\theta}\right)$ rather than a point estimate.

Getting the observed value z_k at k-th step, one updates the distribution of certainty with respect to the parameter $\vec{\theta}$ as $p_k\left(\vec{\theta}\right) \rightarrow p_{k+1}\left(\vec{\theta} \mid z_k\right)$. For this update, one can use the Bayesian formula [1],

$$p_{k+1}\left(\vec{\theta} \mid z_k\right) \propto p\left(z_k \mid \vec{\theta}\right) \cdot p_k\left(\vec{\theta}\right) , \tag{13.1}$$

which is called Bayesian update. ... That is the mundane explanation as seen in many textbooks.

With this explanation, some beginners might misunderstand that the graphical shape of $\left\{p\left(\vec{\theta}_j\right)\right\}_j$ was actually updated in numerical manner on the discrete grid $\left\{\vec{\theta}_j\right\}_j$). They might have misleading impression that the Eq. (13.1) is used to generate update data of $\left\{p_{k+1}\left(\vec{\theta}_j \mid z_k\right)\right\}_j$ composed by taking the product. If one proceed

---

[1] We dare to use z so as not to confuse it with the variable x for the data distribution in the main text Sect. 7.2.3 which cites this appendix.

© The Author(s), under exclusive license to Springer Nature Singapore Pte Ltd. 2023   273
R. Maezono, *Ab initio Calculation Tutorial*,
https://doi.org/10.1007/978-981-99-0919-3_13

with this misleading impression, then one will be at a loss to understand 'what value should be given for $p\left(z_k \mid \vec{\theta}\right)$ reflecting the observed data?', 'what is the distribution of this p in the first place?' etc.

In fact, it is not evaluated numerically as such. Instead of that, the prediction of the parameters are actually updated by an explicit analytical expression like $\vec{\theta}_{k+1} = F\left(z_k, \vec{\theta}_k\right)$, and in order to obtain such an updating formula, Eq. (13.1) is used analytically, not numerically as explained below.

To obtain such an updating formula, we first specify in a little more detail the form of $p_{k+1}\left(\vec{\theta} \mid z\right) \to \vec{\theta}_{k+1}$, namely how to get the prediction of parameters from a given distribution in terms of the choices as L_0 estimate (MAP estimate), L_1 estimate (MED estimate), and L_2 estimation (MMSE estimation) [2]. If we take L_2 estimation for example, the prediction of parameters is performed so that the relation,

$$\frac{\partial}{\partial \theta_{k+1}} \int_{\Omega_\theta} d\theta \cdot p_{k+1}\left(\vec{\theta} \mid z\right) \cdot (\theta_{k+1} - \theta)^2 = 0 \ !, \tag{13.2}$$

is satisfied [2].

The distribution function $p\left(z \mid \vec{\theta}\right)$ is usually set to a simplified model distribution that can be expressed analytically with the parameters $\vec{\theta}$, such as normal distribution or binomial distribution. Substituting Eq. (13.1) into Eq. (13.2) with setting the model distribution function to p, we can perform the integration analytically[2] to get the relation in a form of $\vec{\theta}_{k+1} = F\left(z_k, \vec{\theta}_k\right)$, which can be used to update the estimate of the parameters upon the observed z_k.

13.2   Sparse Modeling by Norm Regularization

Let us remember the elementary setup of data regression, i.e., the problem to construct a model describing data y by N descriptors $\{x_k\}_{k=1}^N$ as

$$y \sim \beta_1 \cdot x_1 + \beta_2 \cdot x_2 + \cdots + \beta_N \cdot x_N = \sum_k \beta_k \cdot x_k , \tag{13.3}$$

by optimizing the fitting parameter $\{\beta_j\}$ by the least-squares method.

Suppose that M properties $\{y_j\}_{j=1}^M$ are each fitted as

$$y_j \sim \sum_k \beta_k \cdot x_k^{(j)} . \tag{13.4}$$

---

[2] It can be said that the functional form of the model distribution is chosen so that this integral can be performed analytically.

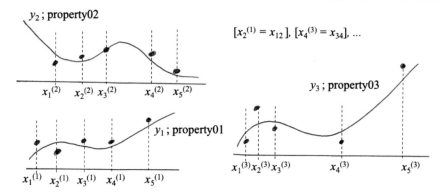

Fig. 13.1 Fitting several properties y_1, y_2, $\cdots$ by each different set of observation points on x-axis. Taking the indexing as $x_j^{(k)} = x_{kj}$, we can formulate it as a matrix-vector product as in Eq. (13.5)

Taking the vector notations, $\{y_j\}_{j=1}^{M} = \vec{y}$ and $\{\beta_k\}_{l=1}^{M} = \vec{\beta}$, and the indexing rule $x_k^{(j)} = x_{jk}$, the fitting is expressed as

$$y_j \sim \sum_k x_{jk} \cdot \beta_k \quad i.e., \quad \vec{y} \sim X \cdot \vec{\beta} , \tag{13.5}$$

in a form of a matrix-vector product. Then, the least-squares problem is expressed as

$$\min_{\vec{\beta}} \left| \vec{y} - X \cdot \vec{\beta} \right|^2 . \tag{13.6}$$

If we try to explain the data with a large number of explanatory variables using only procedures such as the least squares method, we will be forced to generate a sinuous regression curve that passes through all data points shown as $f_2(x)$ in Fig. 13.2. In order to reproduce such a curve, the absolute value of the coefficients $\{\beta_j\}$ in the expansion of the regression curve can be very large, in the thousands or tens of thousands.

However, such explanations using abnormally large expansion coefficients are not convincing. **Regularization** is a way to implement the idea that penalties are given when the expansion coefficient tries to take a large value, so that large expansion coefficients do not occur[3] [3].

To prevent $\vec{\beta}$ from taking large absolute values, we implement it as the minimization of the following quantity,

$$\min_{\vec{\beta}} \left[\left| \vec{y} - X \cdot \vec{\beta} \right|^2 + \lambda \cdot \left| \vec{\beta} \right|^p \right] , \tag{13.7}$$

---

[3] See the explanation in the main text Sect. 7.3.3. The λ in the expression Eq. (13.7) is called the regularization parameter.

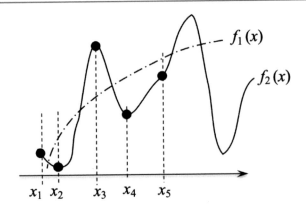

Fig. 13.2 In contrast to the natural fitting $f_1(x)$, the over-trained fitting $f_2(x)$ results in a very large value for the power expansion coefficient, as it is sinuous in such a way as to force all points to pass

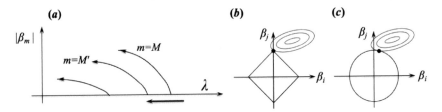

Fig. 13.3 Trimming of descriptors by regularization (dimension reduction of descriptor space) [panel **a**]. Decreasing the intensity of the penalty term λ [horizontal axis in panel **a**], we can rank the significance of descriptors in the order of the first standing β_m and the next standing $\beta_{m'}$. Panel **b** and **c** explains the difference between LASSO [$p = 1$, panel **b**] and the Ridge regularization [$p = 2$, panel **c**]

with the penalty being proportional to the p-power (*e.g.*, $p = 2$) of $\vec{\beta}$. The larger the parameter λ, the more weight to keep the absolute value of $\vec{\beta}$ as small as possible works. As λ gets larger and larger, all the expansion coefficients are suppressed to be zero as a result of the stabilization of $\left|\vec{\beta}\right| = 0$ in this limit.

If there is an mth expansion coefficient β_m that persistently tries to remain finite with increasing λ, then the corresponding explanatory variable x_m can be considered a reasonably significant explanatory variable. Conversely, when λ is gradually reduced from larger values, we can rank the significance of an explanatory variable in the order of the first standing β_m and the next standing $\beta_{m'}$ (Fig. 13.3a). According to this ranking, one can reduce the number of descriptors by picking up the desired number from higher ranking and discarding the rest. This procedure is called 'trimming of descriptors by regularization' (dimension reduction of descriptor space) [4].

For the power p in Eq. (13.7), the choice with $p = 2$ is called Ridge regularization while $p = 1$ is called LASSO regularization.[4] For the reasons described below, it has been clarified around the late 1990s that Ridge regularization does not work well, while LASSO works stably, and then its practical use has greatly progressed [5].

The reason why LASSO regularization works stably is explained as follows. Recalling Lagrange multiplier method, minimization of Eq. (13.7) is equivalent to performing least squares under the constraint $\left|\vec{\beta}\right|^p = $ const. Figure 13.3b, c shows an example with two-dimensional case. In the case of $p = 2$, the minimum solution (x_1, x_2) locates on a circumference where the contour of the first term of Eq. (13.7) takes the smallest value (Fig. 13.3c). In the case of $p = 1$, the solution is located on the edge of the diamond instead of the circumference (Fig. 13.3b).

Compared to the search for the solution on the circumference ($p = 2$), the search using diamond edges often takes the optimal solution on the vertex because the sharp vertex can penetrate more inside the contour line to get lower values. When the vertex is the optimal solution, the solution gives $\beta_m \neq 0$, $\beta_{m' \neq m} = 0$, corresponding to a behavior like Fig. 13.3a, where "only β_m rises to a finite value while other parameters are suppressed. On the other hand, in the case of Ridge regression ($p = 2$), all points on the circumference are equivalent and no such preference to get $\beta_{m' \neq m} = 0$ is taken, being difficult to get the behavior like Fig. 13.3a.

References

1. "Mathematical Methods for Physics and Engineering" Ken F. Riley, Mike P. Hobson, Stephen J. Bence, (Third Edition), Cambridge University Press (2006/6/8) ISBN-13:978-0521683395
2. "Bayesian Optimization and Data Science", Francesco Archetti, Antonio Candelieri, Springer (2019/10/7) ISBN-13:978-3030244934
3. "Sparse Modeling: Theory, Algorithms, and Applications" Irina Rish, Genady Grabarnik, CRC Press (2014/12/1) ISBN-13:978-1439828694
4. "Synergy of Binary Substitution for Improving Cycle Performance in LiNiO2 Revealed by ab initio Materials Informatics", T. Yoshida, R. Maezono, and K. Hongo, ACS Omega 5, 13403 (2020). https://doi.org/10.1021/acsomega.0c01649
5. "The Elements of Statistical Learning: Data Mining, Inference, and Prediction", Trevor Hastie, Robert Tibshirani, Jerome Friedman, Springer (2009/3/1) ISBN-13:978-0387848570

---

[4] Ridge is a name of a person while LASSO stands for 'Least Absolute Shrinkage and Selection Operator'.

Correction to: Ab initio Calculation Tutorial

Correction to:
R. Maezono, *Ab initio Calculation Tutorial*,
https://doi.org/10.1007/978-981-99-0919-3

In the original version of the book, belated corrections have been incorporated in Chapters 2, 3 and 9. The book has been updated with the changes.

The updated version of these chapters can be found at
https://doi.org/10.1007/978-981-99-0919-3_2
https://doi.org/10.1007/978-981-99-0919-3_3
https://doi.org/10.1007/978-981-99-0919-3_9

Printed in the United States
by Baker & Taylor Publisher Services